KU-166-075

the MAGIC AND MYSTERY *of* BIRDS

ALSO BY NOAH STRYCKER

Among Penguins

the MAGIC AND MYSTERY *of* BIRDS

THE SURPRISING LIVES OF BIRDS
AND WHAT THEY REVEAL
ABOUT BEING HUMAN

NOAH STRYCKER

SOUVENIR PRESS

First published in the USA under the title 'The Thing with Feathers'
by Riverhead Books.

Published by arrangement with Riverhead Books,
a member of Penguin Group (USA) LLC

First published in Great Britain in 2014 by Souvenir Press Ltd
43 Great Russell Street, London WC1B 3PD

ISBN 9780285642799

Book design by Nicole Laroche
Bird illustrations by Noah Strycker

Printed and bound in Great Britain by
TJ International Ltd, Padstow, Cornwall

CONTENTS

INTRODUCTION

Imagine what might happen if birds studied *us*.

Which human traits would catch their interest? How would they draw conclusions?

Perhaps birds would begin, as most good scientists do, with the basics. They could spend a lot of time measuring the human body: weight, height, strength, pulse, brain size, lung capacity, color, growth rate, life expectancy, and so on. Academically minded birds could fill volumes with clinical, physical observations of people. Of course, they'd have to dispatch teams of field techs to collect the data. You might step out your front door one morning to find yourself entangled in an invisible net, surrounded by efficient young robins armed with rulers and scales. No doubt they'd soon send you on your way, none the worse for wear save the embarrassment of being caught and the loss of a few carefully plucked strands of hair. Then the robins would retreat to analyze their numbers.

How much could a bird really get to know us from these physical statistics? Take brain size, the feature we humans tend to be most proud of. A bird might rightly point out that for sheer size, the human brain is nothing special; whale and elephant brains, for instance, are much larger. Birds could go a step further and compare human brain size with body weight, but even then, humans wouldn't stand out: Our ratio of brain to body weight is about the same as that of a mouse (about $1/40$) and is smaller than that of some birds ($1/14$); in terms of relative size, ants may have the biggest brains of all (1/7). To explain these mediocre numbers, humans have observed that while brain size increases along with body weight, it follows a power law—in other words, brains don't increase in direct proportion to bodies. But this formula was developed by mammals, for mammals. Birds, if they were to study our brains, might not embrace this logic. They might not think that the human brain—or the human animal, for that matter—was particularly interesting.

To really learn about humans, curious birds would therefore have to study more than just our bodies. They would have to closely observe how we behave, and then try to figure out why we act in the ways we do—a monumental task. Consider the deceptively simple question: Why are you reading this book? You might say "to learn about birds" or "for entertainment," but your reasons probably go beyond that. Reading seems to fulfill a widespread human urge. Evolutionary biologists point out that people of many cultures enjoy reading despite the fact that for most of human history, writing didn't exist. Deciphering words on a page must take advantage of abilities that have been hardwired into us, but nobody quite knows why we like to read. And if *we* don't know why we do it, what could a bird possibly make of you reading this book? How would a bird draw conclusions about a behavior so foreign to its own?

If birds set out to study human behavior, they'd almost certainly start with something familiar to them. They could, for instance, study our sleep habits. That team of robin field techs would camp out in a corner of your bedroom, making detailed notes about things like the color of your bedsheets and the volume of your snores. Surely birds would understand the need to rest each night. But what would they conclude about the wider human condition from these bedroom vigils, especially considering that birds don't sleep the same way we do? Most birds are very light sleepers, rarely falling into the same dead-to-the-world state so familiar to people. And a few birds have particularly strange sleep habits—some swifts are thought to sleep on the wing, perhaps with half their brain turned off at a time. There is a type of parrot that sleeps hanging upside down, like a bat. Hummingbirds enter a state of near-death torpor after dark to conserve energy. Even the most basic behaviors, like sleep, become more complex the more they are studied—and more difficult to understand.

From their research on human habits, birds might even conclude that people wish they were birds. Think of the trillions we have spent on airplanes, space shuttles, and other flying machines over the past century. What is a bird to think? Would birds argue over the differences between bird and human behavior—and whether those differences are absolute or merely a matter of degree, as Darwin once suggested? A bird might examine some of our airplanes with pity, and feel superior to its human study subjects. Who could blame it?

THE IDEA OF BIRDS studying people is pure anthropomorphism—assigning human characteristics to birds just to make a point. Birds have better things to do than study humans, and it's debatable whether they even have the mental capacity to comprehend concepts like scientific inquiry. Birdwatchers often joke about birds watching *them*, but birds probably don't nurture much interest in us beyond a basic fear of predators (check out the chapter "Fight or Flight: What Penguins Are Afraid Of" for more about this). We play only a minor role in the bird world.

And yet the more we humans study birds and discover more

about their behaviors, the more similarities we find between ourselves and our feathered friends. In almost any realm of bird behavior—reproduction, populations, movements, daily rhythms, communication, navigation, intelligence, and so on—there are deep and meaningful parallels with our own. A recent shift in scientific thinking about animal behavior encourages us to concentrate less on the uniqueness of humans and more on what the human animal shares with other animals. Traditionally human characteristics such as dancing to music (see the chapter "Beat Generation: Dancing Parrots and Our Strange Love of Music"), recognizing one's reflection and sense of self (see "Magpie in the Mirror: Reflections on Avian Self-Awareness"), creating art (see "Arts and Craftiness: The Aesthetics of Bowerbird Seduction"), and even love and romance (see "Wandering Hearts: The Tricky Question of Albatross Love") are now recognized in birds. This isn't anthropomorphism at all; anyone who suggests otherwise is ignoring a large part of what it means to be a bird. Moreover, a wave of neurological research on people indicates that the same behaviors, when expressed in humans, may be more instinctive than many of us realize, the result of aeons of natural selection—behaviors that evolved, in other words, because they give us a survival advantage. So the perceived gap between humans and other animals has lately been shrinking at both ends.

I have been fortunate to spend much of the past decade in the field, working on hands-on research projects with scientists studying bird behavior. These projects have allowed me to spend months at a time watching birds in some of the world's most remote places: the Ecuadorian Amazon, a penguin colony in Antarctica, the Australian outback, California's Farallon Islands, Costa Rican and Panamanian jungles, the Galápagos Islands, the Falkland Islands, remote islands in Maine, Hawaii's Big Island, and others. I've observed nearly 2,500 species of birds with the ever-growing realization that they are not our subjects, but rather lively, unpredictable individuals loaded with personality and spirit. It takes time to get to know birds, as it takes time to get to know anyone.

Some bird behaviors don't apply to humans, and those are especially fascinating and exotic: a "sixth" magnetic sense (see "Fly Away Home: How Pigeons Get Around"), flocks that operate as magnets (see "Spontaneous Order: The Curious Magnetism of Starling Flocks"), and the smelling power of turkey vultures (see "The Buzzard's Nostril: Sniffing Out a Turkey Vulture's Talents"). It's hard to imagine having such superpowers, though birds sometimes inspire us to try.

But if you look closely enough, many seemingly incredible bird feats have human counterparts, with interesting lessons. Cooperative nesting in fairy-wrens (see "Fairy Helpers: When Cooperation Is Just a Game") helps illustrate why humans are usually nice to one another. The dazzling speed of hummingbirds (see "Hummingbird Wars: Implications of Flight in the Fast Lane") serves as a warning about our own quickening pace of life. Snowy owls (see "Snow Flurries: Owls, Invasions, and Wanderlust") confirm that not all who wander are lost. Even the domestic chicken (see "Seeing Red: When the Pecking Order Breaks Down") has something to teach us about the natural pecking order.

This book may be about the bird world, but it's also about the human world. Birds can behave in curious, flashy, and startling ways, but they seek the same basic things we do: food, shelter, territory, safety, companionship, a legacy. Each of these chapters explores a compelling bird behavior and focuses on a bird that embodies it. Here you'll read one amazing bird story after another. Prepare to be blown away by, for example, the memory of Clark's nutcrackers (see "Cache Memory: How Nutcrackers Hoard Information"), which show us what the brain is capable of, and might inspire us to boost our own brainpower.

By studying birds, we ultimately learn about ourselves. Bird behavior offers a mirror in which we can reflect on human behavior. In *The Thing with Feathers*, the mirror is all around, glinting from the wingtips of hundreds of billions of the 10,000 species of birds that share this planet with us. Lucky for us, birds are everywhere. All we have to do is watch.

part one
BODY

HOMING

FLOCKS

SMELL

WANDERLUST

PACE OF LIFE

fly away home

HOW PIGEONS GET AROUND

On a recent birding trip, when I stopped at the remote outpost of Fields, in southeast Oregon, to grab a burger, I almost didn't notice the pigeon in the parking lot. Fields is not much more than a store, an inn, and a stand of cottonwoods along an open-range highway in a zip code containing fewer than eighty residents. Several years ago, I watched an airplane land on the highway and taxi right up to the gas pump. The pilot had to be more careful of cows than cars. Fields doesn't get much traffic.

The pigeon was quietly pecking around the pavement and scraps of tumbleweed just outside the station's screen door, looking as though it maybe wanted to come inside. As I finished the last bites of my burger, something clicked. Here, a hundred miles from the nearest McDonald's, any pigeon was a rarity.

"Hey, look, a pigeon!" I said.

Some other travelers also noticed it, and shooed the bird past the gas pump. It hardly moved until they walked right up to it.

"Seems pretty tame," my dad and fellow birder said. "Wonder where it came from."

"I bet we can find out," I replied. "Check out the bands on its legs. I think it's a racing pigeon."

I had recently been researching homing behavior and my head was full of odd tales of transatlantic Manx shearwaters, a wonder dog named Bobbie, and a million-dollar pigeon race in South Africa. Now an actual racing pigeon had dropped out of the sky in the middle of nowhere, straight into my lunch break—a strange coincidence.

I grabbed my binoculars, passed through the screen door,

and began sidestepping in circles around the bird in the parking lot, angling to read its digits. If I could get the full number, I could find out who owned the bird and, possibly, how it ended up all the way out in Fields.

My dad was less subtle.

"Here, help me flank it," he said as he rapidly closed in. The pigeon scooted away at the last second but stopped within a few feet and glanced back with a flirtatious look. My dad charged again, but the wily pigeon zigzagged out of reach. I started to take off my jacket to toss over the bird, but before I could do it, my dad made a bare-handed grab on his third try. The captured bird was nonchalant. It rested comfortably in my dad's hands while staring pigeon-eyed at us, and seemed to expect to be fed.

It looked like a handsome city pigeon with some speckled white feathers on its head, and sported a green band on one leg and a red band on the other. The green one, which held a computer chip, was blank, but the red band was clearly inscribed: AU 2011 IDA 1961.

"Bingo," I said.

After recording the number, we let the bird go free in the parking lot, where it went back to pecking windblown seeds out of cracks in the pavement. Was it lost? Or, like us, just refueling? It would be an interesting mystery to solve. We headed back inside to pay for our burgers.

Birds are amazingly good at navigation, so I thought the pigeon stood a fair chance of making it home. Racing pigeons are deservedly famous for their orienteering skills, but lots of birds have the same ability. I recalled the incredible story I'd heard about an experiment with Manx shearwaters in the 1950s.

"Hey, did you know that a shearwater once covered three thousand two hundred miles across the Atlantic to get home?" I said as we settled the check.

My dad is used to this sort of comment from me, but the waitress gave us a strange look.

JUST BEFORE WORLD WAR II, a Welsh ornithologist, Ronald Lockley, captured two Manx shearwaters, a type of streamlined seabird, on Wales's Skokholm Island, and flew with them by airplane to Venice to try an experiment. On arrival, Lockley walked to the nearest Italian beach and released his two birds. He wondered whether he'd ever see them again.

Fourteen days later, one of them turned up back in its burrow on Skokholm Island, not long after Lockley himself had returned to his home there. He was shocked. The black-and-white, football-sized seabird had traveled more than 930 miles, averaging at least 65 miles per day over mountainous terrain entirely unfamiliar to its kind. Manx shearwaters of this subspecies spend nearly their entire lives at sea, dine exclusively on fish and other marine creatures, and don't normally inhabit the Mediterranean region at all; they reach land only to nest on rugged islands like Skokholm along the fringes of the wild northern Atlantic. A water route from Venice to Skokholm would have required a circuitous 3,700-mile passage southeast around the tip of Italy, westward past Spain and through the Strait of Gibraltar, and northward past Portugal and France, but this bird had apparently taken a more direct flight. Upon release, instead of heading for the open Mediterranean, it oriented in the opposite direction and disappeared *inland*, toward the Italian Alps—and, eventually, arrived home in Wales. Just as though it had a map and a compass.

Lockley was fascinated. He'd settled on Skokholm, a cliff-ringed haven little more than a mile in length, in the 1930s to breed and sell rabbits, but had quickly found a better living writing about the island's birdlife. He'd go on to publish more than fifty books and even won an Oscar for a documentary about gannets—another type of seabird—but is still best known for his experiments with Manx shearwaters and their incredible homing abilities. After the Venice test, he looked for an opportunity to send one even farther afield. A couple of birds packed by steamship to America did not survive the trip in good enough health to return, but when the American clarinetist Rosario Mazzeo visited Skokholm after the war ended, Lockley seized another chance. He convinced his friend to bring two shearwaters home on the plane, to be released in Boston.

Mazzeo's own journey began with an overnight sleeper train from Wales to London. His little carton containing the pair of shearwaters, he later reported, "caused no little wonder and merriment to the people in the adjoining rooms, who could not understand the origin of the mewing and cackling sounds which came from my room in the late evening." The next morning he took a long flight to the United States with the birds tucked under his seat—a trip that would be nearly impossible in today's security age. Only one survived. Mazzeo was met by an airline employee, who escorted him in an official truck to the easternmost edge of Logan International Airport, where they carefully opened the carton and watched its remaining occupant stretch its wings, flutter into the air, and glide away toward Boston Harbor. When the shearwater reached the shore, it abruptly turned east and knifed toward the open Atlantic, where 3,200 miles of ocean separated it from home.

Twelve days, twelve hours, and thirty-one minutes later, Lockley found the shearwater, number AX6587, back in its burrow on Skokholm Island. The seabird had averaged 250 miles per day over the trackless Atlantic for nearly two weeks straight. Mazzeo received a triumphant telegram addressed to Symphony Hall in Boston but didn't get the full story until Lockley worked out the details. When the bird showed up at Skokholm so soon, Lockley was convinced that something had gone wrong; he figured Mazzeo had preemptively freed his shearwater in London. In fact, the friendly clarinetist had mailed a letter from Boston immediately after releasing his charge, but the bird outpaced even the postal service. Only when Mazzeo's letter arrived in Wales, a day after the shearwater, did Lockley realize the incredible trip the bird had taken from America back to its nest in Europe.

THE WORLD ABOUNDS with barely believable stories about animals finding their way home from strange places. Many involve pets. In 1923, a family from Oregon lost their dog, Bobbie, on a car trip to Indiana. After searching exhaustively, they returned home with heavy hearts only to be surprised when, six months later, Bobbie turned up on their doorstep in Oregon—recognizable by three scars and a missing tooth—with worn feet and a mangy coat, skinny and scrawny, having apparently walked 2,600 miles across the country in the dead of winter. Newspapers picked up the story, and Bobbie the Wonder Dog rocketed to instant fame; his family received hundreds of letters, keys to cities, medals, a jewel-studded collar, and a dog-sized bungalow. More than 40,000 people visited him at the Portland Home Show. Bobbie's story was later published in a book, and he subsequently played himself in a silent film, *The Call of the West.* When he died, the mayor of Portland gave his eulogy, Rin Tin Tin laid a wreath on his grave, and his hometown of Silverton initiated an annual pet parade that continues more than eighty years later.

And then there's the case of Ninja, an eight-year-old tiger cat whose family moved from Utah to Washington in 1996. When he was let outside for the first time at his new house in Seattle, Ninja jumped the fence and was never seen again—until more than a year later, when an identical-looking kitty with the same personality and same strange howl turned up at his original house in Utah, looking like he'd just "been through the war," according to the neighbor who discovered him. Coincidence? Or did Ninja walk 850 miles back to his old home? The story was credible enough to be featured on an episode of the TV show *Nature*, along with Sooty, a cat who came back—not quite the very next day—after his family moved more than one hundred miles in England.

Some wild animals seem to have the same instincts. In the 1970s, the National Park Service relocated hundreds of misbehaving black bears in Yosemite, but no matter how far away they carried the sedated bears by helicopter, the animals kept stubbornly reappearing in their old haunts, causing park rangers to give up on relocation and institute a scare-tactic conditioning program instead. Smallmouth black bass, a fish native to eastern North America, have been shown to return to favorite pools after being dumped far from familiar tributaries within their river system. Even snails can find their way home: The typical English garden variety must be relocated more than 300 feet away or it will crawl right back to eat more of your lettuce.

But birds, with their ability to fly long distances and navigate along the way, are exceptional at returning home from unfamiliar places. Lockley's Manx shearwaters are just one example. Even small songbirds can do it. When a group of researchers captured some white-crowned sparrows in Southern California and transported them to Louisiana, many returned to the exact same wintering location in California the following year. The researchers flew some sparrows from California to Maryland, and those came back, too. Not to be outdone, the scientists transported yet another group of sparrows to Seoul, Korea, more than 5,600 miles across the Pacific Ocean from Southern California, where no white-crowned sparrow had ever been recorded. That group never made it home—either the birds fell in love with kimchi or, more likely, some physiological limit had finally been reached.

Pigeons are best known for this ability, partly because of the sport of pigeon racing. Routine pigeon races cover one or two hundred miles, but some official races are longer. In China, there is a race that forces the birds to fly about 1,250 miles (though this distance equates more to survival than fun for the pigeons, making the ethics of this event questionable), and there are anecdotes of pigeons homing successfully from even farther away, on journeys exceeding 2,000 miles. These are incredible feats of navigation, considering the birds have no information about the outward journey before they are stranded far from home.

The homing ability of birds can be positively eerie, which has led generations of researchers and psychologists to wonder whether an ambiguous sixth sense might be involved. In 1898, one Captain Renaud, a French specialist in charge of the military pigeon service, called this the sense of orientation—distinct from sight, hearing, smell, touch, and taste—and attributed it to some organ in the canals of the inner ear. More recently, the controversial biologist Rupert Sheldrake, known for his compelling investigations of telepathy, crystals, and Chinese medicine, has suggested "the existence of a sense of direction as yet unrecognized by institutional science" in birds and other animals. Sheldrake operates on the fringes of science and belief—the journal *Nature* called his first publication "a book for burning"—but the bestselling author attributes his distrust of hard-nosed science to years of keeping homing pigeons as a boy. When he bicycled far from his house to release his pigeons, they'd always beat him home, and scientists couldn't explain how. Years later, Sheldrake continues to ask questions inspired by those birds, and scientists are still working on the answers.

It's easy to see why Renaud, Sheldrake, and others became so mystified by the navigational ability of birds; their sense of direction sometimes seems like magic. But we actually know a lot about how birds navigate. Over the past few decades, researchers have demonstrated that birds can orient themselves based on visual landmarks, the sun, and stars, and even by sense of smell, just like we can. Increasingly sophisticated research now shows that birds are also able to find their way using methods unimaginable to humans, such as magnetic fields, polarized light, echolocation, and infrasound. You can blindfold a bird, cover its nostrils, cover its ears, transport it far from home in a magnetized cage, and, more often than not, it will still manage to find its way home. With so many techniques at their disposal, the question becomes not how birds find their way, but how they ever manage to become lost (a rare occurrence). Yes, we can learn a lot from pesky pigeons.

PIGEONS WERE FIRST DOMESTICATED at least 5,000 years ago, even before chickens, probably near Mesopotamia. Egyptians were apparently training carrier pigeons by 1000 B.C., and some of the world's most influential leaders, including Genghis Khan and Julius Caesar, used them for long-distance communication.

For a while, homing pigeons were most notable for their use in military operations. When Napoleon was defeated at Waterloo in 1815, a swift-flying carrier pigeon delivered the message from present-day Belgium across the English Channel to Count Rothschild, of the Rothschild banking dynasty, who was apparently the first person in England to hear the news. The quick-thinking count made several critical financial decisions and amassed a considerable fortune based on his advance knowledge of the outcome of Napoleon's last campaign.

During the four-month siege of Paris in the Franco-Prussian War, in 1871, the French military used hot-air balloons to transport carrier pigeons over enemy lines, fitting the birds with microfilm that could accommodate hundreds of notes at once. The birds carried more than a million messages into Paris from points as distant as London.

Historians have estimated that half a million pigeons were used by the combined armies of World War I. The U.S. Army Signal Corps used thousands of birds, including one named Cher Ami that saved two hundred U.S. soldiers in 1918 by delivering a message despite a direct shot to the breast that took out one eye and shattered a leg. The heroic bird was later awarded the French Croix de Guerre (War Cross) for her service; she died after retiring to the United States in 1919, and was mounted for display at the Smithsonian Institution. During World War II, the British military alone employed some 250,000 carrier pigeons. Though radio was already widespread, pigeons were ideal for situations requiring radio silence, and they were embraced by both sides of the conflict. Military pigeon programs were eventually disbanded in the 1950s.

The civilian sport of pigeon racing, meanwhile, took off in Belgium in the early nineteenth century when fanciers began to focus on speed and endurance. From there, the pursuit spread to the rest of the world. With the invention of a "rubber countermark" device to measure finishing times in the 1880s, breeders were eager to pit their birds against one another in local races. Military pigeons may have been phased out, but pigeon racing remains as popular as ever. Not much has changed since the early days, except that the stakes are now higher in international events.

Taiwan boasts the most racing events of any single country, with more than half a million Taiwanese enthusiasts participating. Belgium remains a pigeon powerhouse, and racing is popular across most of Europe. The United States has its own racing union, as do most developed countries, with tens of thousands of registered lofts across North America. It's likely that right now, as you read this, pigeons are competing in an event somewhere in the world.

In a typical pigeon race, owners bring their birds to a common starting area, then the pigeons fly to their respective homes—so each bird covers a different distance, depending on its destination. Winners have the highest average speed. Recently, "one-loft" races have gained momentum: young pigeons are shipped to a single location, where they are trained months ahead of time to return to the same loft with all the other entrants so that on race day the group starts and ends together, more like a traditional marathon. Because all birds receive the same training, these races measure the quality of the birds themselves.

Some pigeons have more natural ability than others. Feral city pigeons seem to have a poor sense of direction, as do the doves in pet shops. (The white "doves" sometimes released at weddings and funerals are usually specialized homing pigeons. Untrained birds are apt to get confused, flutter in circles, and fall victim to cats and hawks, not something anyone wants to see at a dignified or romantic ceremony.) Among hundreds of domestic pigeon breeds, most are hopeless navigators; for instance, the Birmingham roller excels at spectacular aerial backflips, and the tippler has extraordinary endurance—the record is twenty-two hours of continuous flight, circling its loft—but neither one is very good at orienting itself. Only one breed, the Racing Homer, is used for serious pigeon racing. Because homing appears to be partly inherited, breeders select the best individuals over many generations. Other birds, like Lockley's shearwaters, also have the characteristic, but pigeons are the only birds trained by us to exploit it. Since we can lure them with food and shelter, pigeons make ideal study subjects, and they have taught us some surprising things.

TO BE ABLE to find their way home from an unfamiliar place, birds must carry a figurative map and compass in their brains. The map tells them where they are, and the compass tells them which direction to fly, even when they are released with no frame of reference to their loft.

Researchers have gone to great lengths to confirm that pigeons don't merely memorize their outward trip. In one experiment, birds were transported in sealed containers filled with purified air, mounted on tilting turntables between coils that varied the magnetic field, and exposed to loud noises and flashes of light, so that, unlike a blindfolded person in the backseat of a taxi who might remember the twists and turns of the journey, they had no external cues. In another study, pigeons were anesthetized and unconscious during the outward trip. They still made it home, proving the existence of an intrinsic map and compass system.

The most basic map is visual. Birds have excellent sight and use landmarks to navigate, just like us. Pigeons learn their local area during short training flights close to home. On longer flights, they have even been tracked following roads, making abrupt ninety-degree turns over intersections. When they are on familiar ground, pigeons use their surroundings as a giant map, just like we do.

It's when the birds are over unfamiliar territory that things get really interesting. Because they don't know the landmarks, they must use other, more refined methods of navigation—the inner compass. Unlike our compasses, birds use several methods to determine direction. If one method doesn't work for some reason, they'll switch to a backup.

Many birds orient based on the sun. In experiments, captive starlings exposed to a movable lamp in place of the sun shift their direction according to the position of the lamp. The sun moves, and birds compensate by the time of day. Starlings that have been artificially conditioned to a different day length, again using indoor lamps, also shift their direction when exposed to real sunshine and orient the wrong way. Pigeons may use the sun as their main compass, but they still make it home on cloudy days—so other, more advanced orientation techniques must kick in.

At night, some birds are able to navigate by the stars, apparently not by using specific constellations but by watching the rotation of the entire sky. When placed in a planetarium, buntings (a type of small songbird) orient themselves based on whatever star the sky revolves around, whether the North Star or a fictitious point. Many songbirds migrate at night, and this might help explain how they navigate after dark. Pigeons are diurnal and don't do as well at night, when they'd rather sleep; they typically rest after the sun goes down and wait for dawn.

For this reason, most pigeon races are held during the day, in clear weather. But the birds do occasionally reach their home loft in the wee hours of the night, confirming that they can fly after dark if they have to.

Cover a pigeon's eyes and it will probably still get home by using its other senses to navigate. One owner fitted his birds with frosted glasses and described them blindly "helicoptering down" out of the sky to reach their loft. In the 1970s, a series of experiments seemed to show that pigeons (as well as a variety of invertebrates and possibly other animals) can sense the linear polarization of sky light and thus interpret the sun's position even on cloudy days, though the importance of polarization remains unclear. Even more interesting, researchers in Europe tried individually covering the eyes of migratory robins and found that the birds could navigate well without using the left eye but got lost when the right eye was blocked. The birds might be using some kind of photomagnetic receptors, processed by the left side of the brain, which has stronger ties with the right eyeball. In other words, the birds could "see" earth's magnetic field, but only in one eye—a bizarre sense that humans can't relate to at all.

There is plenty of evidence that pigeons and other birds use natural magnetic fields just like a traditional compass. How they detect the fields is up for debate—perhaps it's the heavy iron content of cells in their inner ears or specialized receptors in their eyes—but robins have been shown to orient to powerful magnetized coils in captivity, although they can't tell the difference between north and south, and other experiments have confirmed similar sensitivities. Recently, researchers have isolated groups of neurons in the brain stems of pigeons that become active according to the birds' orientations in an artificial magnetic field; the brain stem had been linked previously to activity in the inner ear. Again, we humans don't share this sensitivity.

The map is harder to explain than the compass. How can a bird have a map of a place it has never visited? It must use a coordinate system. Pigeons may cue in to minute variations in the earth's magnetic field to figure out their position on a worldwide grid, but nobody knows for sure. And recent experiments have raised other possibilities.

In one study, researchers isolated the nerves associated with magnetic receptors and sense of smell in the brain. Pigeons made it home just fine when their magnetic nerves were clipped, but they got lost when the nerves relating to smell were cut. Those pigeons apparently used an olfactory map of their environment, literally following their noses to get home. Though birds are usually said to lack a potent sense of smell, this idea is shifting; seabirds, particularly, are now known to find their nest burrows and even recognize their mates by smell alone, and others (like vultures) can pick up on minute concentrations of airborne particles. It's possible that pigeons can sniff their way across the landscape much like a dog does.

In 2011, another group of researchers, in Italy, went a step further when they tested their idea that pigeons may rely on one nostril more than the other while navigating. They inserted rubber plugs into either the right or left nostril of thirty-one pigeons, attached global-positioning-system tags to their backs, and released the birds about twenty-six miles away from their home loft. The birds with their right nostril blocked took significantly more circuitous paths on their homeward journey. Olfactory information is processed in the left hemisphere of the pigeon brain, which connects more strongly to the right nostril, so it makes sense that the right side of their noses would be critical if the birds used smell to navigate. This connection isn't limited to birds, either; though humans, on average, can smell better with the left nostril than the right, we tend to find odors more pleasant when inhaled through the right nostril. The yoga breathing discipline of Swara holds that the right nostril is hot and solar, the left cool and lunar—you decide whether that might carry over to pigeons, but there does seem to be a measurable difference between the two sides of the nose.

The latest research also indicates that pigeons can sense infrasound, the low-frequency noise of the ocean and air currents, and orient themselves accordingly. These sound waves, below the range of human hearing (and near the frequency of earthquakes and whale songs), are so long that they pass through the ground, sometimes traveling hundreds of miles. They also change with atmospheric conditions, so the sounds may travel farther on some days than others. In the 1990s, one geologist speculated that the supersonic boom of the Concorde was interfering with pigeon races because the birds seemed to lose their way more often on clear days when the plane was flying. That was never confirmed, but it's possible; a 2013 study from New York found that homing pigeons got lost on days when infrasound from their home loft didn't reach the point where they were released. In other words, the birds needed to hear the sounds of home to get there.

We tend to think of pigeons as indifferent, consistent, and pretty dumb creatures, hardwired in some mysterious way, but we probably don't give them enough credit for intelligence. Birds are not mechanical missiles. Each one has its own personality, and is apt to make different decisions than another with separate genetics and history. Rather than relying on any one method, pigeons probably use all the tools at their disposal—landmarks, sun, stars, polarized light, magnetic fields, smell, infrasound, and anything else that might help—to return home as best they can while being repeatedly stranded by their owners. These birds are most remarkable in the way they process information, which comes pretty close to the definition of intelligence. But they're not infallible.

Though pigeons are celebrated for their ability to return home from any point on the map, there are a couple of exceptions—places on the landscape that for whatever reason seem to confuse the birds. One location in New York, called Jersey Hill, became famous as a trouble spot for pigeons in Cornell University ornithology experiments during the 1980s. Some believed it was a place of magnetic field anomalies, which would trip up pigeons' magnetic sense. More recently, researchers have mapped infrasound in the area, and concluded that Jersey Hill lies in a sound shadow, a quiet spot where low-frequency sound from the home loft doesn't reach. And a spot in eastern England has become known among pigeon fanciers as the "Birdmuda Triangle" for the number of birds that have disappeared during races there, though the triangle hasn't been studied as intensely as Jersey Hill. Some say that radio signals from a nearby Royal Air Force satellite station jam the pigeons' navigational equipment, but there is no evidence, as yet, that the birds can sense radio signals. For pigeons, certain areas just seem to present more navigational challenges than others.

Perhaps the all-time "most lost" award for a racing pigeon goes to Houdini, who disappeared during a 224-mile race in Britain only to show up five weeks later, in perfect health, on a rooftop in Panama City—5,200 miles away across the Atlantic. Houdini was thought to have hitched a ride on a ship headed for the Panama Canal. "I didn't even know where Panama was," Houdini's owner reportedly said.

Even weirder, on one fateful day in 1998, more than 2,200 pigeons freakishly vanished during two separate races in Virginia and Pennsylvania on the same morning. Sixteen hundred of 1,800 pigeons went missing in one race and 600 of 700 birds never finished the other, amounting to an 85 percent loss rate. The story made national news. Nobody could say where the birds went or even offer a logical explanation. The weather was calm. It's normal for a few pigeons to disappear during a 150-mile race, taken by peregrine falcons or guillotined by power lines, but the casualty rate is usually less than 5 percent; such a mass disappearance was nearly unprecedented. Organizers could only scratch their heads. None of the missing birds were ever found.

New technology might at least give us a clue to the mystery. Tiny global-positioning-system tags developed to be worn by pigeons in miniature backpacks, logging precise tracks of their movements, have suggested that pigeons maintain a hierarchical order in the air. The birds are fairly social, and when released, they usually form a tight flock and travel together back to the loft. Some birds tend to follow the decisions of others, so a few leaders end up guiding the pack, just like chickens in a coop. Social interaction probably helps less-experienced birds learn from older ones, a phenomenon largely ignored by scientists trying to explain the biology of homing behavior. This pecking order is usually a good system, but it isn't perfect. In the case of the disappearing pigeons, perhaps a few leaders became confused—by infrasound, magnetic anomalies, or something else—and many others simply followed them over the horizon, never to return.

Even with all this instinctive know-how, homing pigeons must be intensively trained not to get lost. An owner will typically start by letting his birds fly around their loft, getting a feel for the local area. Then the birds will go on a series of training flights, starting close to home and gradually working farther and farther away. It's not magic. The birds have some innate ability, but they still need conditioning, just like human athletes, to be successful.

THIS TRAINING is on prominent display at the biggest one-loft pigeon event in the world: the South African Million Dollar Pigeon Race. Each January, thousands of pigeon fanciers converge on the glittering Emperors Palace resort in Johannesburg to celebrate their birds' powers of navigation—and a bit of high-roller betting.

At the Million Dollar, for one weekend every year, pigeons become bona fide celebrities, with an entourage to match. Queen Elizabeth II and Mike Tyson have both entered birds in the race. (Tyson once explained that after he quit boxing, keeping pigeons helped him stay sane, and he even starred in his own Animal Planet reality TV show about pigeon racing.) Part of the draw is the ridiculously high stakes. Breeders pay a flat $1,000 entry fee for the chance at $1.3 million in prizes, and winning birds, with names like Rubellos, East of Eden, and Four Starzzz Dream, are subsequently auctioned off for small fortunes as future breeders. In 2008, a particularly athletic pigeon named Birdy went for $102,000.

The race itself is straightforward. More than 3,500 pigeons are released simultaneously from a truck parked about 350 miles from the resort. The first bird to make it back to the resort wins. Pigeons are fitted with electronic chips, the same kind that marathon runners wear, to record their finishing times. The fastest birds usually complete the race in eight or ten hours, depending on weather conditions, while thousands of human spectators watch live coverage on several jumbo screens in an indoor arena.

Everything about the race is extravagant, from its origins at Sun International's Sun City Resort—an over-the-top casino in northern South Africa—to today's incarnation at the Emperors Palace, complete with car giveaways for training runs and flashy auctions. The event is definitely not for the faint of heart or wallet. Elite pigeon racing has been increasingly infiltrated by Chinese magnates and northern European fanciers (a technical term) who can afford birds worth more than luxury cars, to the disgruntlement of some traditional breeders. At its highest level, the sport is all about numbers—dollars and seconds—and it's easy to get mired in endless lists of rankings that form the bulk of the Million Dollar's press releases and reports. But the magic remains. Anyone can understand a pigeon race, and everyone has wondered at some point about the birds that seem to carry a map and compass wherever they go.

The race organizers take training seriously because they don't want prized birds to get lost during the main event, which is longer than most pigeon races. The birds are kept in state-of-the-art, quarantined facilities and pampered like movie stars. In the months before the race, they take part in twenty-seven different training flights, working up from 30 to 240 miles. Because some disappear on each of these shorter flights, owners may enter up to five reserve birds to rotate in. But even with the finest lodging and training, many of the best pigeons in the world still vanish after being dropped 350 miles from home in the remote bushveld of South Africa. No bird is perfect. On race day, the Million Dollar loss rate has varied between 30 and 70 percent.

THE LOST PIGEON IN FIELDS belonged to Marty, who lived in Nampa, Idaho, 110 miles from where I found it. I got his phone number and called him up.

"Oh, yes," he remembered, "she was one of our best birds, nice and speckled-looking."

Marty, forty-four years old, builds manufactured homes during the day and races pigeons in his free time. He was introduced to the sport by his dad, who keeps an aviary with eighty to one hundred pigeons in his backyard; father and son spend nearly every weekend together to train and race their birds. They compete in an Idaho club with about twenty other pigeon owners, racing their birds against one another in half a dozen official events each spring and fall.

This particular pigeon, number 1961 (Marty doesn't name his birds beyond their band numbers), had been released with 150 others on a training toss from Owyhee, Nevada, on April 7, in preparation for a race in that area a few days later. It should have traveled about 110 miles north to reach Marty's loft in Nampa, Idaho, but had somehow ended up 130 miles west, in Fields, Oregon, a week later. Of all the birds released that day, including thirty of Marty's pigeons, it was the only one not to make it home.

Marty wasn't sure how it got lost, especially since its parents were two of his best breeders. Perhaps it had been frightened by a hawk—a pair of red-tailed hawks nesting in a big tree in his yard had lately been scaring the daylights out of his pigeons, though the hawks didn't seem to go after them—and had then become separated from its group and veered off course. Marty had raised number 1961 with forty-eight other young pigeons the previous year and it had already proved its value in several official races, the longest of which originated near Hoodoo, Oregon, about 275 miles west of its loft in Nampa. Some birds can race for five or six years, and this one should have been looking at a successful career.

"It's unusual for a pigeon to go missing," he told me. Two or three times a year, one of his birds ends up in the wrong loft—having followed others home to a different owner—and Marty gets a call from one of his friends. Only once every couple of years does someone call in with a legitimately lost pigeon. Pretty good, considering his birds take part in training flights every other day during racing season. Marty's pigeons usually fly directly home without stopping, but he has seen a few arrive weeks later, so it was still possible that this one could get its bearings.

When I asked how his birds managed to find their way home, Marty didn't mention magnetic fields or specialized neurons. Over decades of tending his flock an hour or two each day, he has come to know them as individuals. "Pigeons are real smart," he replied simply. "People don't realize it, but these birds are very intelligent."

spontaneous order

THE CURIOUS MAGNETISM OF STARLING FLOCKS

In early November 2011, my inbox suddenly exploded with e-mails from friends, relatives, and casual acquaintances, each one linking to the same two-minute online video clip, titled "Murmuration." Unsure what to expect, I clicked on the link.

For the first twenty seconds, the video shows two young women canoeing on a damp evening in Ireland, filming each other with a shaky hand from inside the boat. You begin to wonder what the point is. Then, at 0:23, the view zooms up and out, New Age music fades in, and as the camera adjusts its exposure, you realize that the Irish dusk is full of flying birds—a *lot* of birds, silhouetted in impossibly coordinated patterns from one horizon to the other, blotting out the clouds with their thousands. For the next eighty-one seconds, the birds perform an intricate display of aerial formation, bunching and flattening like a throng of baitfish under attack; you can even hear the whoosh when they make a low, strafing pass overhead. Then the birds fly away and it's over.

The two women, film students Sophie Windsor Clive and Liberty Smith, were working on a graduation project for the London College of Communication when they decided to take their canoe trip. They had no idea that an enormous flock of European starlings habitually roosts in the area where they planned to visit some ruins on a small island. When the women posted their impromptu starling footage online, hardly anyone noticed at first—it averaged ten views per day for the first week. Then *The Huffington Post* linked to it and the video suddenly went viral, garnering 1.05 million views in twenty-four hours.

Over the next few months, more than 10 million people would see it.

If you've ever watched a flock of starlings (poetically called a murmuration, just as crows make a murder and owls a parliament) go to roost, you know why the video mesmerized so many distracted Web surfers. Starlings habitually gather at night, sometimes by the hundreds of thousands in late summer, and in the evening, just before they settle into bed, the birds collectively patrol the airspace above their sleeping quarters, sometimes for half an hour or more. Few birds in the world form such dense, tight flocks. Why starlings do it is an open question. Burning extra energy before drifting off? Guiding incoming stragglers? Keeping an eye out for predators? But there is no dispute that the spectacle is awe-inspiring. The aerial displays resemble dense smoke caught in the grip of an invisible tornado. It's hard not to wonder how the birds stay in such a rapidly changing formation and manage not to bump into one another.

The aesthetics alone are inspiring. New York–based photographer Richard Barnes, best known for his starkly artistic portraits of Unabomber Ted Kaczynski's cabin, released a captivating collection of black-and-white images of starling flocks over Rome in 2005. His photos are carefully framed against urban horizons. Some are simply beautiful, others sinister and Hitchcockian, but all are somehow magnetic (more on that later). In a statement accompanying Barnes's images, author Jonathan Rosen observes, "Part of the fascination of the starlings is the way they seem to be inscribing some sort of language in the air, if only we could read it."

Starling flocks certainly seem to arise from a vibrant, intricately choreographed essence of life itself, a force that defies understanding. How can a hundred thousand birds zip around at 30 miles per hour, each mere inches from the next, and maintain a cohesive flock while constantly shifting direction? The more you think about it, the more it boggles the mind.

Scientists call this *collective behavior*: essentially, a bunch of individuals acting as a crowd. In this particular case, that behavior is self-organizing, which makes starlings interesting, because many things in this universe tend toward disorder; if a glass shatters on the floor, its pieces won't spontaneously rearrange themselves. Logic and experience would seem to imply that groups of starlings must be influenced by some outside force to stay organized, a force that counteracts the second law of thermodynamics—that physical systems tend toward chaos.

But funny things can happen when individuals behave as a group. Other examples of self-organization include snowflake crystals, our economy, the Internet, and even the evolution of life as we know it. Hundreds of scientific papers, theses, and books have tried to incorporate the same concept into such varied topics as language development and traffic jams. Sometimes organization results from many individuals making small, separate decisions—a sort of "ground-up" mechanism, as opposed to the "top-down" approach generally associated with the imposition of order. Starling flocks might be more spontaneous than they seem.

Economist Jeffrey Goldstein took this thinking a step further in 1999 when he attempted to define *emergence*, a loose property by which complex systems form strictly through simple interactions. Emergence applies to termite mounds, hurricanes, art, rock concerts, financial markets, and religion. Although these systems represent vastly different realms, Goldstein believes they share enough similar characteristics to be considered together; he defined emergence as "the arising of novel and coherent structures, patterns and properties during the process of self-organization in complex systems." Starling flocks, which coalesce from the aether, fit this definition perfectly.

Emergence has become a hot catchphrase over the past decade (popular author Steven Johnson wrote a book about it, *Emergence: The Connected Lives of Ants, Brains, Cities, and Software*), but not everyone believes it is entirely useful. One of the idea's most vocal critics is biologist Peter Corning, who published a paper in the journal *Complexity* that called emergence elusive, ambiguous, and "a venerable concept in search of a theory." He pointed out that a game of chess could be considered an emergent system because the complex game results from a few simple rules. But rules don't cause anything; they merely describe relationships. Though chess may seem to be spontaneously organized—a ray of logic in a chaotic universe—that's only because it is affected by two players who are channeling their energies into it; knowing the rules won't help you predict the outcome of the game.

Starlings must obey physical rules to avoid collisions within their flock, no question. But are the birds chess pieces, beholden to some greater force? Or does flock behavior arise from within? If we knew their rules, could we predict the mesmerizing antics of a hundred thousand birds swirling together through the Irish dusk?

IN 1970, BRITISH MATHEMATICIAN John Conway devised a simple exercise that he dubbed the Game of Life. On a grid with no boundaries, some squares are filled in to start the game. Then, depending on circumstances, each square lives or dies from one generation to the next.

Conway set just two rules. If an occupied square has two or

three neighbors (of eight adjacent squares, including diagonals), it will stay occupied; otherwise, it dies. And if an unoccupied square has exactly three neighbors, it will spontaneously become occupied.

People quickly realized that Conway's "game" was quite interesting. It mimics populations. It produces immensely complex patterns. It can form self-replicating structures that kill their parents. It can, theoretically, perform any algorithmic calculation possible on a modern computer within its single grid. In a two-dimensional, cellular sense, the Game of Life offers fascinating parallels to real life, and it is incredibly simple. Depending on the pattern you start with on the infinite checkerboard, vastly different results are possible; some populations die out after many generations, some skyrocket, but most eventually stabilize at some equilibrium. Tiny adjustments have huge consequences. Occupy five neighboring squares in a certain way, and they eventually increase to a population of 116 and stabilize after 1,103 generations. Move a couple of the initial squares over, though, and they stay forever static. Nudge them again, and they become a unified glider that shoots itself across the board into infinite space.

The Game of Life seems to produce spontaneous order; scatter a few random tiles on a checkerboard, and they emerge into beautifully intricate patterns just by obeying simple rules. Instead of descending into chaos, the Game of Life shows that individual cells can organize themselves into complex structures without an overall plan.

It helped that Conway's exercise coincided with the rise of microcomputers that could process thousands of generations in a short time. As the power of computers increased, so did the possibilities of modeling ever more complicated groups—like flocks of starlings.

The first computer model of flocking birds was developed in 1986 by Craig Reynolds, a graphics expert in California. Reynolds had been a technical assistant on the 1981 film *Looker* and worked on Disney's 1982 film *Tron* as a scene programmer. He was frustrated by the challenge of illustrating lifelike swarms of animals. After some tinkering, he discovered that flocks could be simulated using the same principles of Conway's Game of Life: Instead of designating a leader for others to follow, Reynolds decided to set a couple simple ground rules, sit back, and let a flock form itself.

The result was Boids (pronounced the way someone with a stereotypical Brooklyn accent would say "birds"), a simple program with startlingly complex results. Reynolds created graphic images of individual Boid creatures, represented as tiny triangles, and forced them to: (1) avoid collisions at close range, (2) head in the same average direction as neighbors, and (3) avoid becoming separated from the group. These three rules alone—separation, alignment, and cohesion—produced a convincingly lifelike flock of Boids, which twisted around Reynolds's computer screen almost exactly like a group of real starlings in midair.

Reynolds added an obstacle to his model, and watched with fascination as the Boids smoothly split around it and regrouped on the other side. He couldn't believe how realistic the simulation was—it would be perfect for motion pictures. The first movie to showcase the new effect was Tim Burton's 1992 film *Batman Returns*, which featured bats swarming and penguins marching through Gotham City. Boids and its successors would eventually win Reynolds a 1998 Academy of Motion Picture Arts and Sciences Scientific and Technical Award in recognition of his contributions to behavioral animation.

The implications of Boids reached far beyond motion pictures. Because the program used simple decision-making rules to produce results similar to those of the natural world, it was hailed as an advancement of artificial life and paved the way for subsequent attempts to create artificial intelligence. One of the most wonderful aspects of the model was its unpredictability; despite following straightforward rules, it was impossible to predict the trajectory of the flock of Boids more than a few seconds in advance without programming a prescribed path. The Boids behaved like chess pieces in a game with no players.

After centuries of careful study, we know that particles are generally confined by physical laws. We can predict, for instance, how a gas will behave in certain conditions. But the particles of a gas do not interact except in random collisions; if they were to move with behavioral effects on one another, things would get very complicated very quickly. The world's most powerful computer cannot always predict the long-term path of three celestial bodies moving within one another's gravitational fields, even if their initial velocities and directions are finely measured. Neither can we predict the weather more than a few days in advance—there are just too many interacting elements.

And yet Boids elegantly re-created that unpredictability in a simplified, digital environment. The program seemed to mirror real life, but real flocking data would be necessary for scientists to figure out which models best fit reality. Unfortunately, at the time, gathering data on a live starling flock couldn't be done. Nobody knew how to accurately measure the positions, velocities, and trajectories of thousands of birds simultaneously swirling around the evening sky. A breakthrough was needed, and it would eventually come not from the world of biology but from the world of hard physics.

WHEN TENNIS PHENOM SERENA WILLIAMS cracked a backhand to open the third set of her quarterfinal match against Jennifer Capriati in the 2004 U.S. Open, a line judge ruled the ball good. But the chair umpire, who has ultimate authority, thought it had landed out and overruled the call. Williams threw a tantrum and lost the match. Slow-motion replays later showed that she had been cheated out of the point, and that at least two other balls had been called wrongly against her in the same set. The tournament apologized and yanked the umpire, too late to stop a firestorm of indignant players and fans from voicing their protest. The masses called for change. Of nearly 30,000 respondents to an NBC poll, 82 percent supported the use of instant replays.

Video line calling would be a big step for a traditional sport like tennis, but the technology was already in use elsewhere. Replays were first used to officiate American football games in the 1980s. Cricket and rugby matches followed in 2001, and replays debuted in basketball in 2002. Other sports, including field hockey, baseball, and rodeo, soon embraced the technology. Tennis associations had been testing a system called Hawk-Eye, and within a year of Serena's controversial loss, the instant replay was approved for professional tournaments.

Hawk-Eye uses stereoscopic triangulation, a technology that is, in a sense, as old as vision itself. The reason we have depth perception is because of our binocular sight; overlapping images from each eye, slightly offset from each other, are fused together by the brain into one three-dimensional picture. View-Masters and 3-D movies enhance this effect by presenting simultaneous images taken from slightly different angles. Our brain has no trouble aligning the separate perspectives, but programming a computer to do the same thing is tricky. Hawk-Eye takes images from ten different cameras around a tennis court and fuses them into one model; it decides which pixels represent the tennis ball, flying at more than a hundred miles per hour, and predicts where the ball will land relative to a line on the court. Tests showed that the system is accurate to about one millimeter.

At about the same time that Hawk-Eye's unblinking lens began to gaze down from the stands at major tennis tournaments, a group of Italian physicists and statisticians were peering into the sky over Rome, trying to make sense of the enormous flocks of starlings that appeared as if by magic each evening—the same flocks artistically photographed by Richard Barnes. They knew that bird flocks had been simulated with endless generations of computer models, most of them minor tweaks of Craig Reynolds's Boids program from the 1980s, and that nobody had yet compared any of those models with empirical data from a real flock. The best existing set of observations came from a couple dozen fish swimming around a tank, a far cry from thousands of birds in free space.

The Italians, led by physicists Andrea Cavagna and Irene Giardina, recognized a challenge. In 2007, they set up three cameras on a terrace at the Palazzo Massimo, where, more than four centuries earlier, Pope Sixtus V had built a luxurious villa overlooking the imperial Baths of Diocletian, and where, these days, a large flock of starlings likes to roost at an adjacent train station. The cameras were placed fifty meters apart, pointed at the same patch of sky, and set to snap ten simultaneous high-resolution photos per second, using the same stereo photography methods as Hawk-Eye. The researchers started hanging out on the terrace in the evenings, making the museum's security guards nervous, snapping photos whenever a group of starlings swirled through the space in view of their cameras.

The difficulty of capturing starling flocks in a 3-D model boiled down to the "matching problem," which had bottlenecked the entire study of flocks for many years. A computer must accurately match up specks of thousands of overlapping, individual birds between different images, from separate viewpoints and across time. The Italians spent two years developing an algorithm that could analyze photos of starling flocks and match the birds, reporting that it took a combination of "statistical physics, optimization theory, and computer vision techniques." The breakthrough program could process flocks of up to 8,000 birds with nearly 90 percent accuracy, taking up to two hours to churn through a single eight-second clip. When they fed the program with images from their camera setup, the Italian researchers could visualize starling flocks as nobody had ever done before, with quantitative measurements—a real-life leap of knowledge analogous to Helen Hunt and Bill Paxton's heroic (though fictional) tornado measurements in the movie *Twister.*

Starling flocks, it turns out, are thinner than you might expect—more like a floppy pancake than a football. The pancake slides around in various directions, shifting its appearance, but generally stays parallel to the ground and maintains a constant proportional shape, no matter the size of the flock. The density of the flock is higher toward its edges—starlings are more tightly packed at the pancake's fringes than they are in its center. And starling flocks don't have leaders. When a flock turns, birds fly on equal-radius paths; in other words, they each turn on the same curve at the same speed. In a column of marching soldiers, those on the outside of a turn must march faster to maintain their positions. Starlings don't compensate like soldiers do, so birds at the front of the flock end up on the right side after a left turn, those on the right side end up at the back, and those in the back end up on the left side. One potential benefit of this is that no bird must stay in the front position, which, as in a bicycle race, is aerodynamically least efficient and most tiring. Another benefit is that each bird spends the same amount of time on an edge, where there is more risk of being nabbed by a hawk. Because safety from predators is probably a major reason that starlings form flocks at all, any bird forced to spend all of its time on the edge would be less motivated to stay in the group, and the whole arrangement could fall apart.

Andrea Cavagna and his colleagues were just getting started. Now that they had real data to play with, they could begin sorting out how flocks form in the first place. Because of their varied backgrounds in physics and statistics, the researchers could look at starlings mathematically. Cavagna had spent most of his career studying logical theories of glasses and supercooled liquids, far from the muddy realms of biology. Now he wondered: Can starling flocks be described with the same equations used in particle physics?

If you Google "America's most hated bird," all of the top results refer to starlings. Such universal agreement is rare in matters of opinion, but on this everyone seems to concur: Starlings are rats with wings. When a pair of them built a nest under the eaves of my house one spring, I climbed up to it on a ladder and removed the eggs. They never came back.

"What we admire in ourselves we often abhor in our neighbors," writes Jonathan Rosen, and he's got a point. The starling's only real fault is success. In the late 1800s, a pharmacist in New York released a host of European birds into the city in an aesthetic attempt to sculpt the New World according to the Old. Most of those releases, including European robins, chaffinches, common blackbirds, and Eurasian skylarks, quickly died out, unable to adapt to a new environment, but when several dozen European starlings were set free in New York in 1890, the birds thrived like cockroaches and bred like rabbits. Those few individuals quickly multiplied across the continent into a population of 120 million, distinguishing the European starling as about the seventh most abundant bird species in North America today (after the American robin, dark-eyed junco, red-winged blackbird, red-eyed vireo, white-throated sparrow, and yellow-rumped warbler, according to Partners in Flight). Few species have ever spread so fast or multiplied so quickly—except humans.

Starlings are often blamed for the population collapse of

cavity-nesting birds such as the eastern bluebird, but research shows that this probably isn't true. They are also said to bring down airplanes, but I know of only one fatal U.S. commercial airline crash caused by a flock of starlings (geese, ducks, herons, gulls, and cranes are much more dangerous). Starlings do occasionally damage crops, but their effects are slight compared with those of red-winged blackbirds and Canada geese, which eat millions of dollars in grain each year. More bothersome is their poop; as the U.S. Department of Agriculture puts it, "excrement can create a slipping hazard on sidewalks." In 2008, the New York City Transit Authority paid a man $6 million after he slipped on bird poop in a subway station—but that poop actually belonged to pigeons, not starlings.

Perhaps we should take more time to appreciate the starling's merits. Up close, European starlings are feathered in beautifully iridescent layers, with black bodies that glow green or blue, depending on the light. They're perky birds, with quirky personalities, and are excellent vocal mimics; any given starling can imitate about twenty other birds' sounds. They make good pets when raised from chicks, as their imprinting instinct is strong. Mozart kept a starling for three years and taught it to sing bars of his music. When his bird died, the composer buried it in his backyard and wrote a commemorative poem.

Shakespeare, while writing *Henry IV, Part I* in the 1590s, penned a line that may have unwittingly changed the course of avian history: "I'll have a starling shall be taught to speak nothing but 'Mortimer,'" suggests a character named Hotspur, apparently plotting to remind King Henry of Mortimer's imprisonment by training a starling to constantly repeat the name. Three hundred years later, an eccentric pharmacist in New York—one Eugene Schieffelin, president of the American Acclimatization Society, and a big fan of Shakespeare's—is said to have used the line to justify his introduction of starlings to the United States. According to popular lore, Schieffelin tried to import every bird mentioned in the Bard's plays. Direct evidence for this is shaky at best—the druggist would have had to be pretty sharp to spot such an obscure starling reference (much less every reference to the forty-five-plus species of birds mentioned within Shakespeare's plays), and crazy to act solely on it—but there's no doubt that Schieffelin gave America the starling, for one reason or another.

And now they literally flock by the millions. Though a few tens of thousands is normal, particularly large starling flocks have been documented to contain more than 1.5 million birds. One and a half million! Very few animals ever form groups that big. By comparison, the largest known army ant swarms have been counted at around a million individuals, and the world's largest-ever single-artist concert crowd, a 1994 Rod Stewart performance on Copacabana Beach in Rio de Janeiro, Brazil, drew 3.5 million partiers. Yet those totals wilt in comparison to flocks of passenger pigeons, which used to fly around in groups exceeding a billion birds. Congregations of North Atlantic herring have been documented to contain several billion fish in schools measuring 5 cubic kilometers. Even herring have nothing on the Rocky Mountain locust, a type of grasshopper that used to range across the western United States. A single locust swarm was once estimated to weigh 27 million tons, including 12 *trillion* insects. That particular plague, known as Albert's swarm for the Nebraska physician who documented the event in 1875, apparently covered an area the size of California a quarter-mile deep; that's a locust for every grain of sand in 1,800 loaded dump trucks, every second in 400,000 years—the entire evolutionary history of *Homo sapiens*, or every dollar in the U.S. national debt, all in one group. If starlings ever formed a flock that big, it would weigh twice as much as the living human race. (Alas, the Rocky Mountain locust went extinct about thirty years later, and today only North America and Antarctica don't have locusts. Passenger pigeons also went extinct shortly after forming flocks of billions. Maybe we should be worried.)

Still, a million starlings present an impressive sight. A flock of a mere ten thousand is spellbinding as it twists and turns across the evening sky. And starlings, for all the flak they receive, are generally beautiful birds with funky personalities. Even if they have been unnaturally introduced in many places, starlings deserve more respect.

While unwanted starling populations in the United States are booming, the species is crashing in its native haunts across Europe. The Royal Society for the Protection of Birds recently reported that numbers of starlings in the United Kingdom have fallen between 80 and 90 percent in the past thirty years, the largest decline of any British farmland bird. Nobody knows why; perhaps modern farming methods have eliminated insects from the landscape, or the birds have changed their migration patterns to winter elsewhere. Starlings were red-listed in the United Kingdom in 2002 as a species of severe conservation concern, and recent evidence indicates that their population has declined by 40 million there—from an average of fifteen per household garden to just three—since 1979. The flocks that once floated over Manchester have almost all gone.

ITALIAN PHYSICIST ANDREA CAVAGNA became engrossed early in his career with mathematical theories of supercooled liquids. When the temperature of a liquid is decreased to a certain point, it crystallizes, as water turns to ice. Given certain conditions, though, it's possible to maintain a liquid below that critical temperature, and odd things happen to supercooled liquids; for instance, if you cool a liquid fast enough, it becomes a viscous glass that responds extremely slowly even when returned to normal temperatures. Cavagna spent years delving deep into the physics behind these events.

For his Ph.D. research, he studied theoretical physics under the supervision of Giorgio Parisi, one of Italy's most eminent physicists. Parisi is best known for his investigations of spin glasses, disordered magnets with characteristics similar to chemical glasses, such as windows; he's also made significant contributions to particle physics and quantum field theory, and won a range of prestigious awards, including the Boltzmann and Max Planck medals. If anything, Parisi likes to focus on disorder—in magnets, glasses, pure statistical theory, and whatever else he can find.

Cavagna picked up some of his famous adviser's research tastes, including an interest in disordered systems, and continued to study supercooled liquids and glasses after finishing his Ph.D. In 2006, he signed on to a project called StarFLAG, which would change the course of his career.

The ambitious project, overseen by Parisi, was designed to probe the mechanics of starling flocks in order to understand other swarming systems. Teams of scientists from France, Germany, Hungary, the Netherlands, and Italy worked together, each group tackling a different aspect of the issue—computational models, wind tunnel experiments, social theories, and so forth. Cavagna led the group that used stereoscopic photography to record starling flocks in Rome; when they managed to describe a flock of thousands of birds for the first time, he was hooked. Using his training as a theoretical physicist, Cavagna, with his team, cracked a biological phenomenon that had mystified onlookers for generations—even though he had no experience in biology, wasn't what you'd call a birdwatcher, and had hardly conducted a single tangible experiment in his life.

The objectives of StarFLAG went beyond starling flocks. In a sweeping mission statement, the project aimed not only to capture data on bird flocks, but also to use those data to construct new models of collective behavior. Then, the scientists hoped, they would be able to apply their knowledge to other fields, the way that Cavagna applied his understanding of physics to biology.

"Collective movements are a common phenomenon also in human behaviour," the project announced. "We think that [it] is worthwhile to explore the possibility of using the models built for the description of flocking, to describe economic herding behaviours. In this way we hope to get new tools to understand the reasons of social events, e.g. fashions, social dominance." One researcher on the project decided to focus on how people's friends influenced the music they downloaded and the way that they voted. Cavagna didn't know much about the mechanics of fashion trends and market bubbles, but he could contribute his knowledge of physics to try to describe, mathematically, how starlings form cohesive groups. His team of physicists and statisticians got down to work analyzing the data they'd collected from the terrace of the Palazzo Massimo.

They compared the behavior of individual starlings within a flock to the three basic rules used by models dating back to Boids in the 1980s: separation, cohesion, and alignment. Cavagna's group found that starlings avoid collisions, stay at least a wing's length away from one another, and seldom stray far enough from one another to break up the flock—just as the models assumed. Starlings also align with one another, but not quite in the way that flocking models traditionally predicted: Instead of basing directional decisions on birds within a certain distance, each starling uses its nearest seven neighbors to decide which direction to fly in, no matter how far away they are.

This is an important difference. Topological distance—a comparative measurement, like the number of stops on a subway line—appears to be more important than absolute metric distance within a flock. Future models made this adjustment, with good results: When a certain number of nearest neighbors are used instead of those within a certain distance, flocks become less likely to break up, and can expand and contract more easily in response to predators and other fluctuations. And the number seven is particularly interesting. Cavagna mused that even though each starling could probably see more than a dozen individual members of its flock around it, the birds' brainpower is limited to processing seven at a time.

This is a trait that humans may share with starlings. In 1956, scientist George Miller published a fascinating paper, "The Magical Number Seven, Plus or Minus Two," which has become something of a legend; the paper has been cited, at last count, more than 16,000 times in other publications. Miller discussed a variety of experiments that showed an odd psychological convergence on the number seven, not in a black-cat-and-mirrors way, but much more logically.

He described one experiment where people were presented with a screen on which random patterns of dots were flashed for one-fifth of a second. When fewer than seven dots were shown, people were almost always able to correctly count the dots, but they often resorted to imprecise estimates when more than seven dots were flashed. In another test, a psychologist read aloud lists of random items at a rate of one per second, and then asked people to repeat what they'd just heard. No matter what items were being read—words, letters, numbers—people could store about seven unrelated items at a time in their immediate memory, like the seven digits of a phone number. Although these results have been generalized at times past their scientific usefulness—for instance, self-help resources that advise that PowerPoint presentations should have seven main points as a sort of subconscious trick—we do seem to reach certain cognitive limitations near seven items, and starlings may do the same.

Cavagna's team wanted to describe starling flocks with pure physics. They measured the velocities of birds at different positions within a flock and found, as you'd expect, that birds behaved more like those close to them than they did like those farther away. When Cavagna compared these correlation lengths between flocks of different sizes, though, he discovered that they scaled perfectly with the size of the group—birds behaved similarly across longer distances in bigger flocks. He called this a scale-free correlation, and pointed out that this was a feature of critical systems, poised at a tipping point.

Water becomes a critical system when it freezes, and also when it boils. Avalanches are said to reach a critical point at the moment when they break free. Magnets become critical when they spontaneously align from disorder. Perhaps, Cavagna thought, starlings represent a system that forms flocks at a critical point.

Then the team analyzed flight directions within a starling flock. They found a model that correctly predicted order within the flock, and proceeded to demonstrate, in a dense, eighteen-page paper, that it was mathematically equivalent to a well-known model of magnetic systems at critical points called the Heisenberg model, which uses quantum mechanics to describe magnetic orientations. When iron, say, is cooled below a certain temperature, it spontaneously magnetizes. Electrons within the material align their spins below that critical point. Spontaneous magnetism, Cavagna and other physicists argued, was happening in the alignment of flight directions within a flock of starlings. Equations of magnets, it turns out, can describe a starling flock better than biology can.

CAN EQUATIONS REALLY EXPLAIN how starlings perform coordinated flight? Does physics underlie even the most spontaneous, beautiful displays of life on earth? The answer depends, in a sense, on whether you believe math is discovered or invented; whether it's a pervasive force, guiding every action in this universe, or whether logic is imposed by the human brain. History's most ardent philosophers have agonized over the issue, and they are still arguing.

I like to think that life defies physics, and that the beauty of a cartwheeling flock of starlings originates with the birds themselves rather than in a universal law—in the same way that a Renaissance masterpiece may follow specific rules but still takes a real master to produce. As emergence writer Peter Corning pointed out, knowing the rules doesn't always bring a solution any closer.

Andrea Cavagna believes that physics can help us make sense of the natural world, even in ways that may not seem obvious, and his team certainly documented some fascinating observations about flocking behavior in birds. But millions of online viewers reached the same conclusion by instinct and genuine fascination. However you look at it, a murmuration of starlings is absolutely magnetic.

the buzzard's nostril

SNIFFING OUT A TURKEY VULTURE'S TALENTS

One day in high school, I told my parents I wanted to photograph vultures.

"Great," they said. "But how will you get close enough?"

"By putting a dead deer in our yard."

I'd been inspired by an episode of David Attenborough's *The Life of Birds* documentary, the one where the British filmmaker ventures into a Trinidad rainforest carrying a fist-sized hunk of rotting beef. With great flourish, Attenborough hides the meat under a layer of wet leaves on the dark jungle floor and then backs away, muttering lilting phrases such as: "Their beaks are quite adequate for tearing off strips of flesh." Within forty-five minutes, a turkey vulture appears out of nowhere and uncovers the delicacy—quite a trick, for filmmaker and scavenger alike.

If Sir David could lure a vulture with a funky old steak, then how many vultures might a whole deer carcass attract? Just imagine how smelly *that* would be.

My parents were used to this sort of scheme. They weren't sure how or why their son had become so obsessed with birds, but, considering the various other possible vices, had resigned to embrace this one. Their only stipulation was that the dead animal be placed far enough from the house not to stink up the kitchen.

I'd recently acquired my driver's license and had been cruising the local birding hot spots in an old white Volvo sedan. The car was built like a tank and boasted a two-body trunk—or at least, to my discerning eye, a trunk roomy enough for one average deer carcass. I stocked the trunk with gloves and Hefty bags, and began scouring the countryside for flattened deer.

As it happens, a fine roadkill is hard to locate when you really want one. I found myself in competition with Department of Transportation crews assigned to remove carcasses from the state highways, a local wolf sanctuary with permission to collect cadavers for its carnivores, and the odd redneck with a big freezer. Still, there was reason for optimism. Insurance companies estimate that 1.5 million deer collide with cars annually in the United States, and of these nearly 90 percent perish on impact. A corpse would turn up.

Meanwhile, I took a fresh interest in the vultures around me. The state of Oregon in summer abounds with turkey vultures, usually wafting lazily overhead with raised wing tips, rocking slightly while circling a thermal column of rising air. They're stout-looking birds up close, with a dark, smoky cloak of mahogany and black plumage supporting a bare-skinned, creepy pinkish head. The birds began to fascinate me. And the more time I spent stalking roadkill and vultures alike, the more I realized that these scavengers—and the people who have studied them—are far wackier than even I could have imagined.

WHETHER BIRDS POSSESS a sense of smell has been debated for centuries. There's no doubt that most are visually oriented, like us, and experts have historically downplayed any whiff of avian smelling ability. But traditional views are beginning to soften, and turkey vultures are proving to be one of the biggest exceptions to the rule.

Vultures have long been recognized for their special talents at uncovering carrion, their favorite food. Aristotle mused on the ability of vultures—probably European vultures, unrelated to vultures in the New World—to home in on dead animals, suggesting that they must follow their noses. By the early 1800s, most people believed that vultures find their food by sniffing it out. It was an easy assumption, as the carcasses of dead animals assault even our own mediocre noses, but one worth challenging.

In 1826, John James Audubon reached his own conclusions after experimenting with both turkey and black vultures in the eastern United States. The eminent ornithologist hid hunks of rotten meat beneath pieces of paper and placed them near caged vultures. The vultures seemed not to notice. Then he left a deer stuffed with foam, not meat, in an open field. The vultures investigated it. Finally, he hid a rotting deer corpse under some brush, out of sight of the sky. The vultures never found it. Audubon, at odds with accepted wisdom, decided that the birds did not discover food by smell, but by sight alone. His paper, "Account of the Habits of the Turkey Buzzard, Particularly with the View of Exploding the Opinion Generally Entertained of Its Extraordinary Power of Smelling," was met with widespread ridicule. Seven years later, in publications such as London's *Magazine of Natural History*, the discussion had dissolved into personal attacks on the man himself—who at the time had declared his ambition to paint every bird in North America.

In desperation, Audubon wrote pleading letters to his good friend John Bachman, an American naturalist for whom a sparrow and warbler are now named. Bachman was sympathetic; an excellent field man himself, he had strong views about armchair critics. "It has always appeared to me an act of injustice to condemn any man for expressing an opinion on subjects of Natural History," Bachman later wrote, "merely because he had arrived at different conclusions from those who had lived before him."

Bachman agreed to try his own vulture experiment. He figured the question should be simple to resolve. "No one who will read Mr. Audubon's paper," he said, "can deny that if he intended to deceive the world, he certainly chose a subject where detection was easy and certain."

Not everyone disputed Audubon's vulture study, though. A young Charles Darwin, in his travel memoir, *The Voyage of the Beagle*, described his experiences with South American Andean condors about ten years later. "Remembering the experiments of M. Audubon, on the little smelling powers of carrion-hawks, I tried the following experiment," Darwin wrote, describing how he had once, in Chile, tethered several condors in a row to the base of a wall, then walked back and forth in front of them while carrying rotten meat in a paper wrapper. The birds took no notice until he pushed the putrid package against the beak of an old male condor; at that point, the bird tore off the paper and all the condors suddenly began to struggle against their tethers. "Under the same circumstances, it would have been quite impossible to have deceived a dog," wrote Darwin, who concluded that the condors couldn't smell their food.

Rather than generalize his result to all vultures, Darwin was circumspect. He had heard of vultures collecting on rooftops in the West Indies on houses whose human residents had died without discovery—surely a sign that the birds had sniffed the odor of decay. He also knew of an anatomic study on the large olfactory bulb of turkey vultures and the results of several other field experiments. His conclusion was that "the evidence in favor of and against the acute smelling powers of carrion-vultures is singularly balanced." Though Darwin was in his mid-twenties when he experimented with his condors, and just thirty when he published *The Voyage of the Beagle*, his analytic mind was already apparent; in another twenty years, he would pen *On the Origin of Species* and change the world's views on life itself.

At exactly the same time that Darwin was playing with condors in South America, in the winter of 1833–1834, John Bachman was conducting a series of vulture experiments in South Carolina to resolve the smelly argument once and for all.

First, Bachman set out to disprove an odd myth, circulated in newspapers at the time, that a vulture with a punctured eye would tuck its head under its wing and restore its own sight within a few moments. In Bachman's experiment (which wouldn't go over so well today), a captive turkey vulture's eyes were pierced and it predictably went blind. Recognizing an opportunity, Bachman then tested whether the blinded vulture could detect rotten meat by smell alone. When he dangled the rancid flesh of a hare within an inch of the vulture's beak, the bird made no move; the only way to get the vulture to eat was by putting food directly in its mouth. The poor vulture died twenty-four days after its eyes had been poked out, and Bachman was convinced that his experiment had confirmed Audubon's earlier results.

He didn't stop there. Bachman next gathered up carcasses of a hare, a pheasant, and a kestrel, and left them in a heap in his garden along with a wheelbarrow load of slaughter-pen offal. He covered everything under a raised frame and camouflaged it with brush so that the pile was invisible from above but open to the air at ground level. Although the stench of the "dainty mess" became nearly unbearable after twenty-five days, the many vultures that passed overhead never investigated Bachman's garden. Only the neighborhood dogs raided the offering, and Bachman again concluded that the vultures couldn't smell it. A few manipulations seemed to confirm his conclusion: When he uncovered the pile, the vultures arrived. When he hid the pile under thin canvas and scattered a few pieces on top, the vultures ate the obvious food but didn't go for the goodies below. When he tore a small hole in the canvas to show what was underneath, the vultures dug in with gusto.

Finally, the enterprising Bachman devised a more artistic test. On a new canvas, he painted a life-sized depiction of a dead sheep, skinned and cut open with its entrails showing. When the painting was finished, he set it outside to see what the vultures would do.

"No sooner was this picture placed on the ground, than the Vultures observed it, alighted near, walked over it, and some of them commenced tugging at the painting. They seemed much disappointed and surprised, and after having satisfied their curiosity, flew away," he wrote in a scientific paper describing the vulture studies, barely disguising his glee. Bachman repeated his painted experiment more than fifty times with the same result. As his pièce de résistance, he set the painting within ten feet of the heap of camouflaged offal in his garden. The vultures investigated the picture as usual, but departed without discovering the real treat right under their noses.

As far as Bachman was concerned, Audubon had been vindicated: Vultures find their food by sight, not smell. He wouldn't say that vultures *can't* smell, just that they don't use their noses the way dogs and many other animals do in tracking down a meal. When he published his results, nobody could dispute the findings.

It was time for Bachman to shut down his backyard vulture studies. He feared that the vultures "might become offensive to the neighbors." Like Audubon and Darwin, his early vulture experiments spurred him on to greatness: He later founded Newberry College, tended dying soldiers during the Civil War (and had an arm permanently paralyzed by Sherman's army), published Audubon's folio of American mammals, and dedicated his life to Lutheran ministry, living to the ripe old age of eighty-four—no doubt tickled to the end by his memories of that gruesome skinned-sheep painting in his backyard garden.

WORLDWIDE, THERE ARE TWENTY-THREE different species of vultures, divided into two main groups. Old World and New World vultures are quite distinct, but similar in appearance and habits—an example of convergent evolution. Like bird wings and bat wings, they developed independently to accomplish the same function.

Seven vultures live in the New World: Andean and California condors, respectively of South and North America; the clownlike king vulture of South and Central American rainforests; lesser and greater yellow-headed vultures of South America; and the widely distributed black and turkey vultures so familiar to North Americans. Black vultures keep to the eastern half of the continent, while turkey vultures range all over the map, from southern Canada to the tip of Chile and the Falkland Islands.

Of these, the turkey vulture is best known and most reviled. It is a common bird even in areas inhabited by humans, often seen circling overhead as if waiting for something to die (the spiraling flight actually helps them gain altitude in narrow columns of rising air). Audubon, for one, appreciated their movements: "The flight of the Turkey Buzzard is graceful . . . It sails admirably either high or low, with its wings spread beyond the horizontal position, and their tips bent upward by the weight of the body." Close up, turkey vultures are distinctive. They get their name from a superficial resemblance to that familiar dinner-table animal—with bare heads and black bodies—but, to my eye, the vulture is more arresting than any farm fowl. Perhaps it takes an unusually sympathetic view to detect beauty in a vulture, but I think they look crisp and clean. The turkey vulture's scientific name, *Cathartes aura*, literally translates to "purifying breeze"—more household freshener than landfill scavenger.

It's an apt scientific name, as turkey vultures serve the unenviable task of cleaning up our world. Their dirty reputation is undeserved—rather than spreading germs, they gobble them up. To be able to eat animals that have often perished of disease, together with the attendant microbes of rotting flesh, turkey vultures have stomachs of cast iron.

Vulture digestion has lately caught the interest of medical professionals for exactly this reason. Vulture excrement is, amazingly, completely sterile. One delectable fact of life for vultures is that they habitually defecate down their own legs, which serves two practical benefits: It cools them off through evaporation (vultures can't sweat), and the feces help sterilize the bird's legs, which have often just been dragged through bacteria-filled carcasses. Recent evidence indicates that vulture stomachs can process and sterilize anthrax spores without ill effect. They can also eat botulism-infected carcasses and kill the bacteria while their immune systems deal with associated toxins. It's even possible that vultures can disinfect dead rodents containing hantavirus—the virus goes in, but it doesn't come out. The best way that humans have found to deactivate hantavirus, by contrast, is to either soak it in chemical detergent or blast it with at least 115 degrees of heat. If we could figure out exactly how these scavengers handle such serious infections and poisons, perhaps we might find a way to apply that knowledge to humans—with major implications for preventing biological warfare and epidemics. The answer to why vultures have this ability to cleanse and prevent disease probably lies in their remarkably efficient digestion and immune systems, and in the fact that what makes one species sick will not necessarily affect another. Vultures have evolved to survive—thrive, even—on things that would kill us (and many other animals).

All vultures share this gift of a strong stomach, but different vulture species may rely on different methods to find their food—and these species' differences may help explain why the early vulture experiments seemed to provide mixed results. It would be up to later scientists to get to the bottom of this pungent mystery.

ON MARCH 18, 1937, a school building in the oil community of New London, Texas, exploded without warning while classes were in session. Witnesses stated that the walls bulged and that the school's roof lifted before everything crashed down, killing more than 295 students and teachers—ranking the disaster today as the third deadliest in Texas history (after a 1900 hurricane and a 1947 ship explosion). Further investigation showed that the school had used piped-in natural gas for heating, and that a leak had filled the building with a deadly, odorless mixture. When an instructor flipped on an electric sander at about three in the afternoon, the tiny spark caused the whole building to blow.

Within weeks, new legislation decreed that all natural gas in Texas should be mixed with low concentrations of malodorants—extremely smelly chemicals, such as mercaptans, that immediately alert anyone to the presence of gas. It was an easy solution; any leak could then be detected by the human nose with plenty of warning. The practice of adding smelly chemicals to natural gas lines quickly spread worldwide. Although the idea had been around for a while, it took a major disaster to precipitate its implementation.

Mercaptans are a class of organic compounds known mostly for their foul smell. They're toxic in big doses but are typically found in harmless dilution. One particular arrangement called ethyl mercaptan was once named the "smelliest substance" in existence by *The Guinness Book of World Records*; the human nose can detect it at concentrations of less than one part per billion—about a thousand times lower than the threshold for sulfur dioxide (the penetrating smell of pollution and volcanic gas). Mercaptans help give cooked cabbage, onions, flatulence, cheese, bad breath, and feces their fragrant bouquet. Also present in animal blood and brains, mercaptans are released as a carcass decays, helping to give corpses their smell.

It wasn't long before workers at the Union Oil Company noticed a peculiar side effect of the scented gas. Whenever one of their remote lines sprang a leak, a group of turkey vultures would soon collect overhead, evidently detecting the mercaptans percolating in the atmosphere. The workers began looking for circling vultures when tracking down a leak, a technique that is apparently still used today.

This bit of information went unnoticed for several decades by those studying turkey vultures, who tended to have their own noses buried in labs nowhere near remote gas lines. Though more than a hundred years had passed since Audubon and Bachman conducted their classic experiments, it was an open question as to whether vultures discover their food by sight or smell. Stories kept popping up about turkey vultures following their noses. They'd been seen investigating mushrooms and flowers that gave off strong, carrion-like fragrances. Also, a couple of anatomical studies showed that vultures probably had the capacity to smell out their food, even if they didn't use it. One investigation of 108 bird species found that, of the ten birds with the largest olfactory bulbs relative to their own brain size, nine were seabirds (also known for possessing a keen sense of smell). The other was the turkey vulture, which ranked eighth overall.

In the 1960s, Kenneth E. Stager, then the senior curator of ornithology at the Natural History Museum of Los Angeles County, happened to be talking to a Union Oil worker, who relayed the company wisdom about turkey vultures collecting over gas leaks. Despite his interest in vultures, Stager had never heard of any such thing, and the story stuck in his head. When he decided to tackle the question of vulture smell for his Ph.D. thesis, rerunning some of Audubon's experiments along with other tests, he realized that mercaptans—a common denominator of natural gas and carrion—could be a key to the whole puzzle.

Stager was a hands-on, maverick field ornithologist who had literally proved himself under fire. As a foot soldier in the jungles of Burma during World War II, he once came across a couple of ethnic Kachin people who had just killed a beautiful silver pheasant. Stager struck an on-the-spot deal for the bird's skin and sat down to strip out its meat, but before he could finish, Japanese shells began landing in camp—so the enterprising ornithologist jumped in a trench to finish preparing the specimen. This incident led to his quick transfer to a division studying scrub typhus disease, which in some areas of Southeast Asia was killing five times as many troops as weapons were, and he spent the rest of the war happily collecting bird and mammal specimens in remote parts of China. That silver pheasant is still displayed in the Natural History Museum, and Stager went on to participate in scientific bird-collecting expeditions to Australia, Brazil, Africa, India, Mexico, and various far-flung islands, adding considerably to our current knowledge of the birdlife of those areas.

Inspired to study vultures at home in California, Stager brought his usual straightforward intensity and commitment to the project. He made an exhaustive literature search on vulture smell, still the most encompassing review anyone has ever published on the subject. Then he designed experiments to demonstrate exactly where previous studies had gone wrong.

First, Stager repeated Audubon's basic test of concealing a carcass to see whether turkey vultures would find it. He baited cage traps with carrion and hid them, always out of sight of the sky. When Stager checked his traps, he'd caught some vultures—which must have followed the reek of decay to discover the bait.

Then he hid a carcass and watched where vultures appeared in the sky. The birds tended to fly low, as you might expect if they were searching for ground scents, and usually approached the bait from downwind. He figured, by crude visual estimation, that they could detect the hidden carcass from at least a couple hundred yards away.

But Stager's most interesting test involved the odor of natural gas spiked with mercaptans. Spurred by the stories of oil workers, he devised an ingenious experiment to attract turkey vultures with scent alone. The results were clear: When Stager puffed pure foul-smelling gas across the hills of California, vultures materialized and circled overhead. Interestingly, the shy birds wouldn't land without a decoy carcass placed nearby. Stager reported that in his odor-only trials, turkey vultures would circle for about twenty minutes before wandering away, which suggested their need for visual confirmation of the source of the smell. He concluded that turkey vultures are able to use their noses to find food, even if the final approach is often visual.

Stager and many others have accordingly speculated about flaws in Audubon's and Bachman's early experiments. Audubon conducted many of his tests with black vultures, and the current wisdom is that only turkey vultures possess a good sense of smell—black vultures don't (they often follow turkey vultures to find carcasses). Audubon may have misinterpreted his results by assuming that all vultures act the same. Also, he seemed to believe that vultures like their meat rotten, which is definitely not the case—we now know turkey vultures prefer fresh carcasses—so perhaps the birds smelled Audubon's hidden corpses but just didn't want to eat them. And when Bachman covered his pile of garden offal with canvas to see whether the vultures could find it, he didn't factor in how weak the birds' beaks and claws are; they might have sniffed the food but couldn't reach it through the thick canvas.

Modern experiments have added substantial evidence to support the idea that turkey vultures possess an extraordinary sense of smell. In one study, a researcher painstakingly set out seventy-four chicken carcasses in the understory of a Panamanian rainforest (he purchased freshly slaughtered, fully feathered chickens from a Panama City market, then hustled out to the jungle before they could rot) and monitored how long it took for turkey vultures to discover each one. They found seventy-one of the dead chickens within several days, with no measurable difference between those left on the open forest floor and those covered loosely with leaves. A curious finding was that the birds were more attracted to two-day-old than either one- or four-day-old carcasses. The researcher hypothesized that his chickens weren't ripe enough the first day—not emitting enough gases, including mercaptans, to be smelled from a distance—but that by the fourth day they were too putrid for even a turkey vulture's taste.

If turkey vultures smell their way to food, we still don't know exactly what they sniff that attracts them to a decaying carcass. Two scientists at Humboldt State University tested turkey vultures' reactions to mercaptans to see whether the Union Oil Company workers' story could be verified experimentally—and whether, as Ken Stager once hypothesized, mercaptans are the key to the whole puzzle. Vultures get excited when they smell food, so the researchers exposed captive vultures to increasing concentrations of the gas while measuring the birds' heart rates; when the heart rate spiked, a threshold had been reached. The lab test, somewhat disappointingly, showed that vultures tended to react only when mercaptan concentrations reached about one part per million—impressive for a bird, but not even close to our own ability to detect the same chemical at one part per *billion*.

When the scientists then modeled that concentration of gas dispersing from a single point, they calculated that a turkey vulture would have to fly within seventeen centimeters of the ground to smell a decaying carcass. Either the calculations were incorrect (as some critics have suggested) or turkey vultures are sniffing something else. Mercaptan isn't the only gas released from decaying carcasses; the same study also tried butyric acid (the odor emitted by rancid butter) and trimethylamine (a fishy odor), with similar results. Rotting meat gives off a whole suite of delectable compounds, including such aptly named chemicals as cadaverine and putrescine. Which of these are most inhaled—and enjoyed—by turkey vultures, nobody has any idea.

Studies of turkey vultures' sense of smell are complicated by the fact that smell overlaps with taste, and even turkey vultures are fairly discriminating about what they eat. They prefer herbivores to carnivores, just as we do, which is why you'll see vultures more often on a road-killed deer carcass than, say, a dog or cat. And they like their meat fresh.

Birds, like humans, can distinguish among sweet, sour, salty, and bitter foods, though the arrangement is slightly different. Their taste buds are located mostly on the interior of their beaks instead of on their tongues, and birds have vastly fewer receptors than people do; humans have about 9,000 taste buds, while birds have only a couple hundred. It's unclear how the number of taste buds relates to actual perception (one species of catfish, for instance, has about 100,000), so we don't really know how different birds experience taste. But because smell is so closely related to taste, it's possible that vultures taste their food more intensely than other birds in the world—oh, the irony.

IN HIGH SCHOOL, as my interest in birds was taking off, it took me more than a month to track down a worthy roadkill to lure my turkey vultures. By the time I found a good carcass, I'd nearly lost hope; all the turkey vultures would be gone in a few more weeks, and I wondered if I'd have to put off my photography adventure until the following summer.

But one afternoon when I wasn't even particularly searching, there it was, pummeled and piled on the shoulder of the interstate on a hot August afternoon. A doe had wandered into freeway traffic to meet her unfortunate demise, and, in a circle-of-life kind of way, a new beginning. She looked crumpled, somehow smaller without the use of those spindly legs, and I nearly missed seeing the carcass at all.

I pulled over, popped the trunk, whipped out my gloves, and wrestled the deer off the ground—not an easy task for one skinny kid. Eighteen-wheelers whizzed past, blasting me with heat waves rising from the tarred pavement. The deer was fresh but already bloated with gas, carpeted in flies and yellowjackets, and it oozed a viscous fluid from a nasty shoulder injury. It smelled awful. I trusted the Hefty bags would contain the leak, dropped the carcass in the trunk, jumped in the driver's seat, and headed out.

Even at 65 miles per hour, the overpowering stench could not be contained. It built to unbreathable proportions before I was halfway home, so I drove the rest of the way with my head hanging outside the window like an excited puppy, my emotions suddenly mixed. Was this really a good idea?

One thing that nobody—not Audubon, not Bachman, not Stager—ever mentioned in their scientific papers about vulture smell was, well, the *smell*. This doe of mine reeked to high heaven and beyond. The scent of decay is not subtle. I didn't need to calculate a concentration threshold or gas distribution to confirm the obvious. My nose could easily sniff out this deer from hundreds of yards away.

At home, I'd already set up a camouflaged blind in an overgrown pasture behind my house where sunlight could illuminate the carcass. As soon as I pulled into the driveway, I transferred the deer to a waiting wheelbarrow and rushed it to the spot, dumping the body as if it had naturally been flattened there. Then I settled in to wait.

It didn't take long, not even a few hours. Around dinnertime, a vulture glided in like a stealth bomber and landed in a tree at the edge of my backyard, flapping awkwardly as it perched on a high branch. Several more showed up over the next hour. I scooted out to my makeshift blind and sat next to the stinking deer until the sun set, but none of the vultures ventured down to touch it. Disappointed, I retreated to the house and went to bed.

The next morning, a creepy sight greeted my family outside the kitchen window: Nearly twenty turkey vultures were hunched on top of telephone poles, rooftops, and tree branches around the yard, evidently having spent the previous night roosting nearby. The birds were silent but focused. And they were hungry.

I couldn't believe it. I rushed outside just in time to catch two vultures on the deer carcass, gouging out the deer's eyeballs and gums. They spooked when I climbed into my blind, but forty-five minutes later they were back with a dozen friends, poking their heads into every orifice right in front of my camera lens.

I still remember the success of that feast like it was yesterday. Twenty turkey vultures and a few ravens cleaned that deer down to bare bones in less than a week. Each morning, I found about a dozen of them roosting hunchbacked on top of the house and nearby utility poles, loafing with ghoulish preoccupation as they digested between feedings. The deer, at the center of a trampled area of grass, became the greatest bird feeder I'd ever seen.

When the whole animal had been finally cleaned up, the vultures departed as silently as they'd arrived, leaving me with a series of suitably gory mealtime photographs. After a very long time, the streams of whitewashed guano plastering the yard and the smell in the kitchen also disappeared (as did the strained grins of my parents). But my vulture enthusiasm lingers.

In retrospect, I just wish I'd slid a Hefty bag *under* the deer carcass before dropping it in the trunk. That smell may be heaven to a vulture, but I never quite got used to it. The essence of my precious deer lived for months in the upholstery of my car, long after all those vultures had sensibly migrated south for the winter. At least the scent reminded me, every time I went for a drive, that my gruesome friends would soon be back. When the days started getting longer, and the Oregon rain began to slack off, I monitored the sky with new hope—optimistically searching for the first turkey vulture of spring.

snow flurries

OWLS, INVASIONS, AND WANDERLUST

Just after seven a.m. on October 28, 2011, a commuting birder in northeast Minnesota was startled to notice a snowy owl perched on a lamppost, like a ghostly apparition, next to the bridge that connects Duluth to Superior, Wisconsin. He called his wife, who relayed the sighting on a local rare-bird alert. Three days later, on Halloween, another snowy owl eerily materialized in southwest Minnesota. Then several more popped up across the Midwest. Birders all over North America began to take note. Was this the beginning of an invasion?

By late November, hundreds of snowy owls were lighting up birding hotlines from Oregon to New Jersey and as far south as Kansas, causing expensive traffic jams of cameras and spotting scopes wherever the owls appeared. There could be no doubt: The winter of 2011–2012 was shaping up to be one of the biggest snowy owl irruptions—or population shifts—in history.

On Thanksgiving Day, staff at Honolulu International Airport discovered a white owl sitting on the airfield—the nearest approximation to tundra in the middle of the Pacific Ocean—and, never having dealt with such an obstacle, promptly shot the bird in the name of aircraft safety, never mind that it was the first wild snowy owl ever to have appeared in Hawaii. At Logan International Airport in Boston, where a few owls had landed in past winters, officials were more levelheaded. Inundated by more than forty snowies, they painstakingly captured each one, marked the birds' heads with colorful paint to keep track of them, and moved them to safer locations.

As December and January unfolded, thousands more snowy

owls drifted south. In Missouri, where the previous record high count had been eight, fifty-five were discovered. Kansas had one hundred sixty. One birder in South Dakota ticked twenty of them in seven hours, while observers in Vancouver, British Columbia, counted thirty-one in a single spot at Boundary Bay. A Dallas police officer who casually mentioned to a friend that he'd been watching a snowy owl on a light fixture at his local marina caused a stampede of chasers; it was the first one in Texas in fifty years.

Nobody forgets their first snowy owl sighting. To see one, two feet tall, nearly pure white, with piercing yellow eyes, a wild incarnation of Harry Potter's pet Hedwig, is to gaze into the very soul of the Arctic. Most of us rarely get the chance, because snowies usually stay out of sight of civilization, tucked in the narrow latitudinal belt between the Arctic Ocean and boreal treeline from northern Europe to Russia and Canada, in some of the harshest terrain on earth. They spend most of their lives in the chilly, windswept tundra, where a steady supply of lemmings—small, cute rodents related to voles, which occupy the tundra, eat grass, and reproduce a lot—feeds generation after generation of owls.

Every few winters, though, snowies show up south of their normal range, sometimes in big numbers. In 1916, for instance, more than 1,000 snowies were reported in Washington alone. Many of them never left. Birders at the time carried shotguns instead of binoculars, and a lot of snowy owl specimens from that year can still be found in local attics and museums. Because they stand out and tend to cause excitement when they appear, movements of snowies have been noted for as long as people have been paying attention to birds; particularly large North American irruptions occurred in the winters of 1947–1948, 1966–1967, 1973–1974, 1984–1985, and 1996–1997.

Snowy owl invasions are said to reflect a grim cycle. The owls, according to classic biological literature, are driven out of their homes by food shortages. When northern lemming populations crash every few years, owls are forced to flee south, and most of the birds that reach the United States are starving. It's a textbook example of the delicate predator–prey balance.

It can also be a sad story. A newspaper article, "Snowy Owl Invasion Ends in Tragedy," published in March 2012 after most of the owls had disappeared, warned that the birds hadn't fared well: "99.9 percent that come this far never make it home," one rehabilitator lectured.

Another expert quoted in *The New York Times* agreed. "These birds are starving to death," he said. "No question." It didn't help that according to the narrative, many of the owls were being harassed by birders and photographers who had no qualms about approaching dangerously close to the birds. Bad news for the owls. Their tragedy presented a sad finale to an otherwise compelling story.

At least it would have, if the tragedy were true.

THAT DECEMBER, I got a tip about a possible snowy owl sighting at Fern Ridge Reservoir just west of Eugene, Oregon. Half a dozen snowies had already been reported in the state, which in many years sees none. One celebrated bird behind a retirement village in Albany, Oregon, had received hundreds of visitors, but I hadn't gone to see it. I decided to chase down the more isolated Eugene sighting instead.

Fern Ridge is a large, shallow lake, drawn down in winter for flood control. In December, most of its area is covered by a mudflat dotted by stumps of trees logged in the 1930s. The mudflat is three miles wide, nearly perfectly flat, and resembles a moonscape—or, possibly, to a snowy owl's discerning eye, Arctic tundra. Peregrine falcons, bald eagles, and other raptors like to perch on the stumps, but a local birder had just photographed a round white blob out there that suggested a snowy. Unfortunately, the area is a major pain to access.

The only way to get close enough for a definitive identification was to slog across the mudflat. I packed my rubber boots, spotting scope, and camera, and drove to Fern Ridge on a chilly Monday morning.

The place was deserted. From the edge of the lake bed, my naked eye couldn't detect anything except mud, mud, and more mud, stretching nearly to the horizon. When I peered closer through my high-powered spotting scope, though, I noticed a white smudge on a distant stump. It was too far away to be sure, even magnified sixty times, but it could have been a snowy owl. Or a milk jug. I locked up the car, put on my boots, shouldered my scope and camera, and walked into the moonscape.

It was immediately obvious why the original observer hadn't attempted to get closer. Mud curled around the soles of my boots, and I made slow, squelching progress. As I worked my way farther toward the center of the lake bed, the ground became softer and my feet sank six inches on every step. A couple of times, when my boot stuck in the muck, I tripped and fell forward, burying my outstretched glove in ooze to save my camera. Discarded boat motors, tires, and beer bottles lay scattered among the stumps, half buried in black silt. In summer, windsurfers and yachts would have been whipping around above my head. Now, I heaved my legs across the deserted, sloppy mud, sweating under my jacket in the 40-degree weather.

Every so often I stopped to glance at the distant white blob, but it remained veiled by ground shimmer, sitting stationary.

For all my efforts, I was hardly moving, either. I couldn't tell for sure whether it was a snowy owl or something else.

It took more than an hour to slog close enough for the speck to resolve into something recognizable. By then I was a mile from my car and completely surrounded by a barren expanse of silt. Nothing moved except for a small flock of dunlin, small shorebirds that periodically wheeled around the mudflat on a cold breeze.

But the effort paid off. When I finally set up my spotting scope and peered through it, I gasped. There, filling the view, with the yellow eyes of Arctic summer and a white body of northern winter, was an immaculate snowy owl. It stared with fierce intensity and a bit of curiosity. We sized each other up for a few minutes, the owl occupying its favorite stump and my boots slowly sinking up to their brims in mud, while I snapped a couple of photos. The owl was nearly entirely white, indicating that it was probably an adult male. When I zoomed in to see if my images were sharp, I noticed some bloodstains on the bird's white chest.

"You left a bit of lunch on your chin," I remarked. The owl closed its eyes and sank into a snooze.

I wondered what it had been eating. We were a long way from the nearest lemming—a snowy owl's usual snack—and I doubted any mice or voles were hiding in the muck that surrounded us. Could it catch one of those dunlins zipping around in a confusing swirl, or did it waft out over the nearby lake under cover of darkness to snatch ducks off the water? The owl wasn't telling, and, besides the bloodstains, didn't offer any evidence. It rested, sphinxlike, unmoving except for a couple of loose feathers that ruffled in the breeze.

I retraced my steps to my car, each one repooling into dirty sludge, and contemplated this snowy owl's welfare. The bird didn't seem to be starving, exactly. I hoped it would stick around for other birders to enjoy.

But it was gone the next day. Other birders made the exhausting trek only to find the owl had disappeared. Some snowy owls, like the one in Albany, remained for weeks or even months, staking out winter territories in farm fields, suburbs, and rural areas across the northern United States. Those were the ones that appeared in newspapers and on TV. The Fern Ridge owl was just ghosting through.

SNOWY OWL IRRUPTIONS have been happening as long as anyone can remember. The bird's distinctive white body, yellow eyes, and round frame appear in prehistoric European cave paintings alongside other animals, probably the oldest human art depicting an identifiable bird.

The species was classified by Linnaeus himself, the father of modern scientific names, in 1758. Recent DNA research has indicated that snowy owls are closely related to the great horned owls of the Americas, the several eagle owl species of the Old World, and other members of the genus *Bubo*; though snowies appear to have a rounded head, they have tiny ear tufts, which are usually hidden.

They are the heaviest of North America's owls. This was put on dramatic display about a month after I trekked out to Fern Ridge, when a photographer captured an encounter between a peregrine falcon and a visiting snowy owl in one of Chicago's city parks. The owl, sitting on the ground, suddenly found itself being repeatedly dive-bombed by the falcon, which lived in a nearby neighborhood. The falcon probably just wanted to drive out a perceived competitor, but it's not inconceivable that a peregrine might eat a large, juicy owl if given the chance; I once watched a hungry peregrine falcon annihilate a smaller burrowing owl in California, and peregrines have been documented eating short-eared owls—though great horned owls, in turn, have killed adult peregrines. For its part, the defensive snowy owl became quite agitated, puffed out its feathers, and anticipated each strafing with a little hop into the air and a backward somersault, throwing up its talons for protection as the falcon veered off. After five minutes of sparring, with both birds shrieking and hissing at full volume, the peregrine gave up. The photographer called it a draw.

In a historical context, the 2011–2012 irruption event didn't break many overall records. Washington had closer to 100 sightings than the 1,000 of 1916, and that's with magnitudes more human observers than existed a hundred years ago. Sure, the individual snowy owls in Hawaii and Texas got a lot of attention, but even these sightings weren't all that surprising, because snowies are nomadic birds, prone to wandering.

Some suggested that the Hawaii bird had hopped a ship to get there, but it is more likely to have flown 2,000 miles over the Pacific Ocean under its own power. Not many ships go directly between the Arctic and Hawaii, and snowy owls have been known to wander far from land. In 2012, for instance, one also appeared on Shemya Island, halfway between mainland Alaska and Japan. Over the years, they have been recorded in every U.S. state except Arizona and New Mexico, often along shorelines.

The cool thing about the 2012 irruption was that it was so well documented. Thanks to eBird, a website where birders can log their sightings on one comprehensive database, people anywhere can look at the snowy owl invasion on their own computer screen. For the first time, with just a few clicks, it's possible to view most snowy owl records on a map and sort them by date. This sort of control enables finer analysis from data collected all across the country by birders and garners more interesting results.

If we look at the map, it's clear that most of the snowy owls that came south in 2012 were concentrated in the middle of the United States, with especially impressive numbers in the Great Plains. Though the Pacific Northwest and the Northeast had above-average numbers, you don't have to go back very far to find years with more owls in those states. In the Midwest, by contrast, snowy owls were *everywhere*. The 2012 irruption in that area may have been the biggest anyone's ever seen. Thanks to eBird, we know that the invasion varied regionally. But that still doesn't explain how it happened.

SNOWY OWLS AREN'T THE ONLY BIRDS to make periodic winter irruptions. Type the words *bird irruption* into Google, and you'll get blizzarded with reports of redpolls, grosbeaks, crossbills, nuthatches, chickadees, and waxwings, among others.

All of these birds have something in common: They live in the far north or high mountains. And every few winters, large numbers of them show up in lower, more southern areas, outside their normal range.

These winter irruptions are generally thought to arise from food shortages. Redpolls, for instance, forage on catkins; when the catkin crop fails, they hightail it south in search of something to eat. Nuthatches and crossbills depend on conifer cones, and pine grosbeaks and waxwings follow fruit. In some years, when catkins, cones, and fruit fail in the same season, many species of northern finches and other birds that depend on those food sources simultaneously invade southern areas they don't normally inhabit. When that happens, birders call it a superflight.

Irrupting birds generally have something else in common: They are food specialists. Unlike birds farther south, northern birds live in an area with few meal options—so they specialize. Redpolls are fine eating birch catkins as long as catkins are available, and catkins are usually abundant. When birches have a bad year, though, redpolls are forced to move elsewhere. Specialists like this are typically more nomadic than omnivorous birds because they have to keep up with a fickle food supply. This is also true of some tropical birds that specialize in particular fruits in the rainforest; like northern finches, they are known to wander widely in years when the fruit crop plummets.

Several northern raptors show similar periodic influxes, including rough-legged hawks, northern goshawks, great gray owls, boreal owls, and, of course, snowy owls, but their irruptions are less well understood than the movements of finches and other seed-eating birds. Perhaps the same effect carries upward so that plant-eating mammals, the main diet of Arctic raptors, impose boom-and-bust cycles on predators, or maybe other factors are at work. Things are more complex at the top of the food chain.

AS FOR SNOWY OWLS, the more we learn about them, the more we know we don't know.

In 1945, ornithologist V. E. Shelford published a short paper suggesting that snowy owls fly south in years following population crashes of their main Arctic food source, the collared lemming. Sometimes lemmings breed so fast that their population skyrockets, and then they disperse in search of more sparsely occupied territory. This phenomenon has led to the false but widespread notion that lemmings commit mass suicide by jumping blindly over cliffs, but it also means that their populations are somewhat cyclic. Shelford gathered data on lemming and snowy owl numbers, plotted them on a graph together, and believed he detected a pattern.

It seemed logical. After lemming populations crashed, snowy owls would have less to eat and might have to head south in search of food. Shelford's paper, which had cobbled together data from several different sources, didn't spend much space elaborating on specifics. The idea was easy for anyone to understand, quickly accepted, and used as a textbook example of population dynamics for decades. Many sources still suggest that snowy owl irruptions are caused by lemming shortages, which, according to one study, occur, on average, once every 3.9 years.

The trouble is that there's little hard evidence to support Shelford's theory. It's difficult to prove that invasions follow any kind of cycle; different people look at graphs of snowy owl occurrences in the continental United States and see different things. Because humans love patterns, it's hard not to try to find regularly spaced spikes. People who argue for a snowy owl cycle generally say that irruptions occur every three to five years—but sometimes six; that invasions are sometimes, but not always, followed the next year by a smaller "echo"; that sometimes cycles are skipped altogether; and that really big invasions usually occur every fifteen years—but sometimes ten, and sometimes twenty, and occasionally back-to-back. Looking at a chart of snowy owl sightings over the past couple of decades, my eye gets the impression of regular intervals, but separating the waves from the chop is tough. One serious statistical analysis was unable to quantify any kind of pattern: Peak snowy owl years, at least at the scale studied, could not be predicted at all.

Also, lemming numbers do not seem to follow similar trends across wide swaths of tundra. Populations in one region may experience a peak while, a hundred miles away, their neighbors are crashing. Even Shelford's original paper pointed out that some of his lemming data showed different trends in different areas. Recent research has indicated that lemming populations are probably more of a patchwork than a unified, continent-wide force.

Shelford seemed to assume that snowy owls normally stay in one home area, but snowies move around a lot. They are highly nomadic—probably looking for a ready food supply—and many shift south even in non-invasion winters. Areas in the northern Great Plains and New England host snowy owls every single year. This makes it difficult to define what, exactly, constitutes an irruption. How many is abnormal? Concentrations of owls shift within their wintering range from year to year. Some winters there are more snowy owls in the Northeast; sometimes they're more heavily concentrated in the Northwest. In 2012, as eBird showed us, they hit the Midwest hard.

Owl numbers do appear to be at least partly correlated with lemming numbers. After all, a healthy adult owl needs about five lemmings a day just to stay alive (which makes you wonder: Does a snowy owl wake up in the morning and think, "Yes! A lemming for breakfast! And brunch! And lunch! And two for dinner!"). If irruptions are caused by fluctuations in lemming populations, though, they may be more regional than Shelford believed. And it might not be quite for the reason he described—that lemming crashes cause snowies to move south.

The 2012 snowy owl invasion occurred after a season when

lemming populations were at all-time highs in many parts of the Arctic. That summer, owls had fared exceptionally well, rearing up to seven or eight chicks in an average nest. Normally, it's closer to two. Suddenly all those young owls found themselves jostling for space. Territory became a limiting factor, so some of them headed south. Rather than too *few* lemmings, it seems, there were too *many* owls.

If that's true, then the implications are slightly different for the owls seen in the Lower 48. They're not all starving, just looking for space.

Despite widespread assumptions about starving snowy owls, many or most of them seem to be in good health. In 2012, out of thousands of snowy observations in the United States, I could find only a handful of references to birds dropping dead from starvation. Sadly, human-related injuries are much more common; one ornithologist in Kansas examined five dead snowy owls, of which three were hit by cars, one collided with a train, and one was electrocuted on a utility line. Nearby, one was shot by a poacher, and another hit a power line and broke its wing on the concrete below. Five owls in Nebraska were found with broken wings after collisions with vehicles, and two more were electrocuted there. One in Ontario had its leg wedged in a telephone pole insulator and was photographed dangling from the top of the post like a white plastic grocery bag. Of five snowy owls that were found dead in Massachusetts, all of them were trauma victims. In their usual Arctic home, snowy owls don't have to worry about such things.

A comprehensive study in Alberta found the most common causes of snowy owl mortality to be automobiles, electrocution, gunshot wounds, and collisions with unknown objects, in that order. Only 14 percent of known deaths were attributed to starvation, and most of the owls there make it through the winter without dying. More than half of all examined specimens in Alberta (most of which met traumatic ends) had healthy fat deposits, indicating that they were not only eating well but also had stored extra reserves of energy.

What exactly do they eat, then? Wintering snowy owls, with no access to lemmings, have a surprisingly varied palate. In the Aleutian Islands, one study documented snowy owls surviving on ancient murrelets, a type of small seabird. In the Shetland Islands, the owls eat mostly rabbits. Alberta birds focus on mice and voles. Pellets examined in Oregon show mostly black rat remains. One study of pellets from snowy owls wintering on rocky islands in British Columbia found that the owls were surviving entirely on birds, with more than twenty prey species identified; most were ducks and grebes, but there were also significant numbers of gulls, a few shorebirds, and even a short-eared owl. A particularly aggressive snowy owl was once observed carrying off a full-grown feral cat at John F. Kennedy International Airport in New York. In 2010, a snowy owl in Manitoba made off with a mini–Yorkshire terrier while the dog's owner still held the other end of its leash; fortunately, the quick-thinking owner yanked his disappearing terrier back from the owl's talons, and, though the dog had gone into convulsions from shock, it was unharmed. Snowy owls have extraordinary sight and hearing, are fierce hunters, and will seemingly dine on whatever is easy to catch. Females tend to take bigger prey than males.

It is natural to assume that wintering snowy owls are wasting away because they spend most of their day sitting and are often quite approachable. An adult snowy has few predators besides humans, so they can seem very tame, and they have little regard for the dangers of traffic—they are known for their habit of perching on highway shoulders, barely noticing the vehicles speeding past. Wintering snowy owls spend 98 percent of their time sitting still during the day and may do most of their hunting after dark. In the far north, of course, owls must hunt during summer daylight and winter darkness alike.

Studying snowy owls is difficult because of their tendency to roam. One researcher staked out a population of 1,000 snowies for an Arctic summer only to find, frustratingly, that none of them were present the following year. Another researcher banded seven snowy owl chicks on Victoria Island in northern Canada, hoping to figure out where they ended up. Amazingly, three of them were relocated within seven months: one at Attawapiskat, Ontario (1,350 miles away); one at Clyde Forks, Ontario (1,950 miles away); and one at Sakhalin, Russia (3,450 miles away). The last one had traversed half of Arctic North America and crossed the brutal Bering Sea.

So we can theorize about the causes of irruptions, but their reasons remain as mysterious as the owls themselves. Nobody really knows why snowy owls sometimes appear south of their normal range. Maybe it's because of crashes in lemming populations, maybe it's because of peaks in lemming populations, and maybe it's something else entirely. Because lemmings are so patchy, perhaps a more widespread effect—like weather—might better explain large-scale owl irruption events. In any case, the owls that occasionally show up south of Canada are in better shape than some news reports suggest.

In 2012, the *Weekly World News* announced its own theory. The self-proclaimed "world's only reliable" news source reported that "hostile snow owls" were swarming the United States and "working with alien forces to attack American citizens." According to the article, the owls were communicating with Gootans that landed on planet earth in November 2011 and who were also, incidentally, killing Peruvian dolphins.

Who can say otherwise? Next time you see a snowy owl, better keep a safe distance. Just in case.

"NOT ALL THOSE WHO WANDER ARE LOST," J. R. R. Tolkien once wrote. Like many nuggets of bumper-sticker wisdom, the quip contains an element of scientific truth—and it is especially apt for snowy owls. These owls have perfected their nomadic lifestyle not as a stepping-stone to stability, but as an ultimate survival strategy. The snowy owl "appears to have reached a terminal point in its evolution," author Karel Voous observed, suggesting that the bird's drifting habits are the perfect response to an unpredictable Arctic environment. Snowies wander with a purpose.

Only recently have we begun to appreciate quite how much these birds meander. New satellite technology has radically changed our knowledge of the birds' movements. When transmitters were first attached to several adult female snowy owls at their nests in Barrow, Alaska, in 1999, researchers were amazed to find that the owls took off in different directions after the nesting season, covering up to 1,960 miles to reach northern Siberia and Victoria Island, Canada. None of the owls returned to nest in Barrow the next year; instead, they spent the following summer between 390 and 1,200 miles away. Several crossed the Bering Sea multiple times, even lingering on sea ice for weeks on end. And none of them seemed to follow a regular migratory pattern. Though the owls sometimes returned to places they'd visited before, they didn't commute between summer and winter ranges so much as wander an enormous swath of the Arctic, apparently looking for things to eat.

This study inspired others to track snowy owls. In 2008, one study in northern Canada found, unexpectedly, that six adult female snowy owls spent almost the entire winter on sea ice, far from land. Lemmings don't exist on frozen ocean, so the scientists concluded that the owls were targeting seabirds, such as eiders—a type of bulky duck—instead of small mammals. Snowy owls may not depend on lemmings as much as once thought; even in more usual tundra habitats, they have been documented preying on birds, foxes, and other animals. Their willingness to fly long distances over water and to live on sea ice suggests a certain previously unrecognized adaptability.

The king of snowy owl trackers is Norman Smith, director of a wildlife museum in Massachusetts, who, since 1981, has captured nearly five hundred individual snowies at Boston's Logan Airport. Each winter, when snowy owls turn up on the tundralike airfield at Logan, Smith is called in to safely relocate the birds. He began applying satellite tags to owls in 2000, and the results are fascinating: Many of the snowy owls wintering in Boston return to the high Arctic, from northern Quebec across Hudson Bay to Canada's Nunavut Province, in good health after spending their winter so far south. Most of them don't return to Massachusetts, though a few do. Smith caught one owl at Logan sixteen years after he first banded it there, setting a record for the oldest known wild snowy owl. (Satellite transmitter batteries last only one to three years, so it's hard to know how long the owls live; in captivity, the record is twenty-eight years, and wild snowies might reach that age, too.)

Additional satellite studies in Europe and the North American Arctic have confirmed that healthy adult snowy owls habitually wander thousands of miles, which sheds new light on the birds that show up south of Canada. In the life of a snowy owl, it's no big deal to fly that far; it's just that they don't usually fly that far south, where we can see them.

Such wanderlust is rare in a wild animal, but a few humans

can probably relate. Who hasn't dreamed of disappearing over the horizon? For snowy owls, pulling up stakes isn't a vague fantasy—it's how they survive.

The word *wanderlust* can be traced to apprentices in medieval Germany who journeyed from town to town, gaining practical skills before becoming master craftsmen. A similar tradition still exists in France, where young *compagnons* travel the country to work in communal houses, in what's known as the Tour de France (a pursuit that preceded the unrelated Tour de France bicycle race). Many world cultures have developed traveling coming-of-age traditions, from Australian Aboriginal walkabouts to Native American vision quests and Amish Rumspringa, which could be compared to young snowy owls' gadding about after leaving the nest. But the wandering instinct goes deeper with snowy owls. They seem to be deeply nomadic by nature—more nomadic than humans, with some notable exceptions.

In 1995, a cover article in *New Internationalist* magazine estimated that between 30 and 40 million people in the world are nomads, and most of them are shepherds or herders (virtually all traditional hunter-gatherers have succumbed to modern ways). Like Arabian Bedouin herders, Mongolian tribes, and African Tuaregs, they keep no permanent homes, preferring to stay on the move. The author noted that most nomadic people "live in marginal areas like deserts, steppes and tundra, where mobility becomes a logical and efficient strategy for harvesting scarce resources spread unevenly across wide territories." He could easily have been talking about snowy owls. In the same environments, it seems that humans and Arctic owls have adopted the same survival strategy.

Some people seem to be more inclined to wander than others. This trait could be coded in our genes, and may date back to our distant ancestors. Genetic evidence indicates that modern humans left their African home to colonize the world, starting between 338,000 and 60,000 years ago. Why did those first people go? Were they more adventurous than the ones who stayed behind? Perhaps restlessness has a genetic component. If so, emigrants would be expected to establish populations with more wanderlust in their DNA than those back at home. Scientists have identified one particular allele, called 7R, of our DRD4 gene that may fit this description; it has been linked to attention deficit disorder and attraction to novelty, earning its nickname: the risk gene. Research has documented that people with the 7R allele take 25 percent more financial risk than those without it. Tellingly, the allele tends to be more concentrated in recently established populations (in terms of historic human expansion): Most people in the Americas have it, a few in Europe do, and it is rare in parts of Asia. People with this "wanderlust gene" may be literally hardwired to seek new experiences.

Do snowy owls have a wanderlust gene? They seem to feel the imperative to move as a powerful force, and it's likely to be driven by instinct, honed by aeons of following fickle food supplies. Maybe someday we'll know exactly what drives them on. In the meantime, whenever one of these ghosts of the white Arctic materializes someplace new, all we can do is appreciate the visit—because soon enough the owl will drift back over the horizon, leaving only fleeting memories behind.

hummingbird wars

IMPLICATIONS OF FLIGHT IN THE FAST LANE

Liz Jones, proprietor of the Bosque del Río Tigre Sanctuary and Lodge in Costa Rica, has given up on feeding hummingbirds near her house.

"We put up our first sugar-water feeders about ten years ago," she explained as we sweated one morning in the tropical heat outside the lodge's lush front entrance. "It took several months for the birds to discover the feeders, but when they did, they were quite active. We had nine different species of hummingbirds making regular visits, and many of them nested in our garden."

I could imagine how awesome this must have been. Because the feeders were in plain sight of the outdoor dining area, guests could watch the action as they ate their meals. Birders delighted in seeing their first-ever bronzy hermits, charming hummingbirds, long-billed starthroats, violet-crowned woodnymphs, and white-necked jacobins as they sipped their wine each evening.

Things went well with the feeder setup for a couple of years, until a feisty rufous-tailed hummingbird arrived. Four inches long and the weight of a nickel, he was handsome enough, with an iridescent green body, red lance of a beak, and orange-brick tail. But he was also meaner than all the other hummingbirds and never let anyone forget it. When he wasn't gorging himself on sugar, the aggressive hummer spent almost all his time chasing everyone else away; this guy was the tiniest bully Liz had ever seen.

She tried moving the feeder, but he just moved with it. She tried putting out more feeders so he couldn't possibly guard them all, but another rufous-tailed showed up and they joined forces to defend against all comers, in all corners of the garden. The yard rang constantly with the sound of miniature aerial dogfights. She tried taking down all her feeders but one, thinking that the rufous-tailed would be overwhelmed by the other hummingbirds flocking to a single spot, but that just made it easier for him to fend them off. She even tried putting a feeder inside the lodge. A shy long-billed hermit learned to dart indoors for quick sips, but, lamentably, the other tropical hummingbirds couldn't figure it out.

Pretty soon, the other hummingbirds stopped visiting entirely, leaving the rufous-tailed to sit, hour after hour, next to a lonely feeder. Doubtless, he enjoyed his life, king of an unlimited supply of food, but he wasn't very entertaining for visiting birders.

"It was boring," Liz said after she'd told me the whole story. "The rufous-tailed chased the others away, so instead of a buzzing hummer setup, we eventually just had this one bird, day after day."

After five years, several more rufous-tailed hummingbirds arrived. Between them, they wouldn't let a single other hummer anywhere near the lodge, much less the feeders. Where several species used to nest among the flowers in the garden, now there was just one.

Liz tired of pampering her resident bullies. She took down the feeders and never tried attracting hummingbirds again. By eliminating the feeder setup, she hoped to return to a more balanced mix of hummingbirds around the lodge, like there had been before the feeders went up. Some of the shyer hummers did return, but only gradually.

It was all very vexing. The feeders supplied unlimited nectar, and when they were nearly empty, Liz always refilled them. If the hummingbirds could just get along, they could eat all they wanted without wasting energy on fighting. Why were they so selfish? Hummingbirds are supposedly smart—they may have the largest brain size, relative to body mass, of any bird in the world—but they weren't acting logically. It didn't make sense.

BECAUSE THEY ARE SO TINY, hummingbirds are often described in superlatives. They are the world's smallest birds and the world's smallest warm-blooded vertebrates (besides a couple of obscure species of shrew). The bee hummingbird, which lives in Cuba, weighs as little as 1.8 grams—about a third as much as a sheet of printer paper. You could mail sixteen of them for the price of a single postage stamp.

Most hummers aren't quite that teeny, but they're all small. Of about 330 species of hummingbirds living between Alaska and Chile, the largest, the giant hummingbird of high-elevation Andes woodlands, could still be covered by that same first-class stamp. Not that airmailing hummingbirds would give them much of an edge; the ruby-throated hummingbird routinely flies more than five hundred miles nonstop across the Gulf of Mexico during its spring and fall migrations, taking about twenty hours to do so, and the rufous hummingbird, which commutes annually between Mexico and Alaska, makes the longest flight, relative to body length, of any bird in the world.

Being small carries great advantages. Hummingbirds are nearly immune to predators. Because bite-sized hummers are so quick and light, they don't have to worry much about hawks and other attackers. One study in 1985 could find only thirteen confirmed instances of predation on adult hummingbirds in North America ever, of which several events were classified as "bizarre"—praying mantises, spiders, fish, frogs—and concluded that North American hummingbirds "do not have natural predators in the usual sense."

With this in mind, the authors analyzed hummingbird life spans. Using an equation of body mass and scaling factors to predict life expectancy, they suggested that a hypothetical hummingbird weighing three or four grams should live between 5.5 and 6.1 years, barring disease, predation, or accident. Though nobody really knows how long hummingbirds live, one study in Colorado documented several wild broad-tailed hummingbirds surviving at least eight years, past their theoretical physiological limits, and a broad-tailed currently holds the hummingbird longevity record at about twelve years. Without significant predators, hummingbirds may expect to live into old age as long as they stay healthy. They're just too small and fast for predators to bother with.

The only published account of any animal consistently attacking hummingbirds involves a pair of bat falcons in Venezuela. While watching the small, nimble falcons for 164 days, one researcher observed them catch all kinds of fast-flying prey, including ten species of hummingbirds, two species of swallows, eight species of swifts, and four species of bats. He estimated that the pair of falcons had consumed roughly 600 birds and bats during his study, of which 100 were hummingbirds. Because bat falcons are relatively common and widespread in the tropics of Central and South America, they may have a real effect on hummingbird populations in those areas, but this appears to be the only exception to the no-predators rule.

Another benefit of small size is increased agility. Hummingbirds are the only birds that fly backward, and they have no trouble tracing lines better suited for helicopters than airplanes. Their dexterity means that they can access food sources usually unavailable to other birds, such as hanging flowers, and zip around at 30 miles per hour without poking one another's eyes out.

This agility has lately inspired researchers at the Pentagon, who recently announced the hummingbird-like Nano Drone, a miniature flying robot with two wings that looks impressively like the real thing. The tiny drone, operated by remote control, can hover, maneuver backward and sideways, and dart in and out of buildings while transmitting a live video feed. The military envisions it as a spy device capable of reconnaissance missions and perching near targets without arousing suspicion, though one hopes that birders would be able to spot the difference. Prototypes have even been modeled and painted to closely resemble hummingbirds.

At the moment, the main issue with the drone is its battery life. Early versions could fly for twenty seconds, which has since increased to eight minutes—still not quite long enough to be useful for real-world applications. One of the project's managers has enthusiastically explained that the Nano uses biomimicry to copy natural flight, but he didn't mention the basic problem with living, breathing hummingbirds: They burn energy like fighter jets. To accurately mimic nature, the high-performance drone would have to spend most of its day refueling instead of spying.

In terms of energy, hummingbirds live at the edge of physical possibility. Birds, and other warm-blooded animals, constantly lose energy as heat is transferred through the skin (touch your face and you can feel the warmth leaking out of you). This energy, of course, is supplied by consumed calories. Small animals lose heat much more quickly because their surface area has a greater ratio to their volume—the same reason that small ice cubes melt faster than big ones—so hummers have to eat more relative to their body weight than other birds just to stay warm. If they were any smaller, it would not be possible to eat enough each day to make up for the lost energy.

Large animals have the opposite problem: They can't lose heat fast enough. That's why desert rabbits have big, floppy ears and why camels have long, spindly legs—to increase their surface area, and dump extra heat. Cold-blooded creatures aren't as constrained by heat loss, which is why many of the world's animals smaller than a hummingbird, from butterflies to frogs, do not maintain a constant body temperature.

The extreme lifestyle of hummingbirds is dictated by energy. Hummingbirds have made every possible physical concession to maintain their supercharged existence. To drive such a high metabolism, they have the largest hearts, relative to body mass, of any bird, and the fastest heartbeat of any animal—a hummingbird heart has been measured at more than 1,200 beats per minute in flight, about six times faster than a human's maximum. Their circulatory and respiratory systems are incredibly efficient, from lungs that can process more than 250 breaths per minute to a remarkably high concentration of red blood cells to carry oxygen to their muscles. Because of the metabolic demands of hovering flight, hummingbirds use more oxygen than almost any other animal.

All of this requires a lot of fuel. Hummingbirds routinely ingest more than their own weight in nectar each day, equal to somewhere between three and seven calories—the equivalent of a human eating a couple hundred pounds of hamburger between breakfast and dinner—and pass about 80 percent of it through their kidneys, as if you or I urinated twenty gallons a day!

But that's still not enough. To keep running at normal speed, a hummingbird would have to refuel so often that it would never be able to sleep. So hummers have adapted to shutting off their engines at night, powering down to a state of torpor near death. A sleeping hummingbird lowers its internal body temperature so far that signs of life barely register; its heart slows to a lethargic pace, its breath is difficult to detect, and its metabolism is reduced by as much as 95 percent. While in this state of suspended animation, the hummingbird can't be awakened. A couple of hours before dawn, something flickers in its brain, sending the signal to begin a bout of shivering. Within about twenty minutes, the hummingbird revs back to life, and after a quiet period near sunrise, it zooms off in search of breakfast.

AS ANYONE WHO FEEDS THEM KNOWS, hummingbirds are not the least bit cuddly. Backyard birders are sometimes horrified to watch their hummers body-slamming, clawing, and tearing at one another in knockdown fights around the yard.

"I thought they were just cute little interesting birds and had this sweet image of them in my mind," a concerned blogger once wrote. "It was a shock for me to witness such unexpected violence."

A birder in Alabama has complained, "Sitting on my back porch is like sitting in a Lilliputian battle field with miniature helicopters humming overhead," while a gardener in Maine observed, "Everyone seems to love hummingbirds except other hummingbirds. One has to wonder why feeders have multiple perches. Their very nature keeps them at war with one another."

The Aztecs, who knew a bit about violence, figured this out long ago and named a hummingbird as their god of war: Huitzilopochtli, roughly translated as the "hummingbird on the left," who demanded occasional human sacrifices to stave off the end of the world. He was usually depicted with a feathered head and was said to be so bright that soldiers could see him only by peering through the arrow slits in their shields. When Aztec warriors died in battle, they were believed to return to earth as hummingbirds.

I experienced the true ferocity of hummingbirds one afternoon in the highlands of Costa Rica. While walking along a remote gravel road above San José, I noticed something wriggling in the ditch next to my feet, bent down, and casually closed the fingers of my right hand over two fiery-throated hummingbirds locked together.

They didn't even notice me until I had scooped them up. Apparently, the birds had been so preoccupied with fighting each other, struggling in a shimmery heap in the bottom of the ditch, that they were oblivious to the rest of the world. Now I regarded the heads of two tiny Aztec warriors scowling out from between the fingers of my fist, and wondered what to do with them.

Hummingbirds don't have the weaponry to inflict serious damage on each other. Their beaks are softer and more sensitive than they appear, so stories of hummers spearing one another through the gut are probably exaggerated. Besides, long beaks are ineffective at close range—any wrestler could tell you not to bring a javelin to hand-to-hand combat.

Those beaks are surprisingly bendy, more like a springy diving board than a javelin. This flexibility wasn't recognized until 2004, when researcher Margaret Rubega analyzed ultra-high-speed videos of hummingbirds catching fruit flies in mid-air, and demonstrated that their lower jaw can flex as much as 25 degrees even though it is made of solid bone and lacks a hinging joint present in other bird beaks. This setup allows hummers to open their beaks wider while they pursue tiny flying insects, and, when the bill re-straightens, to snap it shut in less than a hundredth of a second—the same physical mechanism used by Venus flytraps.

Hummer feet are likewise too weak to cause injury. They can barely grasp the branch they sit on, and a perched hummer can't even turn 180 degrees without lifting off to change position, much less walk across flat ground. But that doesn't stop the birds from getting into constant spats.

The only way to defuse a fight is to separate the fighters, so I carefully untangled my two fiery-throated hummingbirds, held one in each hand as they glared at each other from behind my respective thumbnails, and then released them in opposite directions, hoping they'd settled their differences. At least my sudden appearance had interrupted their scuffle.

Hummingbirds generally lead solitary lives, coming into contact with one another only near food sources. Mating is dispensed with quickly; females aren't allowed in male territories except during a brief period in spring, and males of most species don't help build nests, incubate eggs, or care for the young. That's why you'll never see a flock of hummingbirds. Males of a few tropical hummers congregate during the mating season, perching near one another in the dense forest understory to sing for the attention of females, but, for the most part, the birds are stubbornly unsociable.

A study of migrant rufous hummingbirds in California's Sierra Nevada once tried to determine just how selfish the birds were. Researchers wondered whether the hummers would stop defending flowers after eating their fill of nectar or would aggressively keep eating as much as they could cram in all day long. Were hummingbirds "time minimizers," who would consume just as much nectar as they needed, or "energy maximizers," who would eat as much as possible?

At first, the results seemed to show that migrant rufous hummingbirds were time minimizers, because the birds spent about 75 percent of the day perched. But after some analysis, the researchers concluded that the opposite was true: Hummingbirds could fit only so much nectar in their crops before they became too heavy to fly, so most of their sitting time was necessary for digestion. Meanwhile, the hummers continued to hotly defend their flower patches so that they could keep tanking up whenever their stomachs had enough space. They were energy maximizers.

The researchers decided to experiment. They covered half of the flowers in each hummingbird's territory with clear plastic bags, halving the available nectar. In response, the hummingbirds almost universally doubled the size of their territories to include neighboring patches of unclaimed flowers. The birds had to spend more time defending the bigger territories, and it took longer for them to commute from flower to flower while feeding. When the bags were removed, the territories shrank back to normal size.

This supports the idea that hummingbird territories are based on energy—like most things in a hummer's life—rather than physical area, as is true of many animals. The birds know how many flowers they need to control to stay full all the time.

But evidently that logic goes only so far. Give a hummingbird a nice, compact feeder with a continuous supply of nectar, like Liz did at her lodge in Costa Rica, and he'll maximize the opportunity by defending it at all costs. And who can blame him? A hummingbird who finds a feeder must feel like baby Zeus did after he accidentally broke off the horn of his goat nursemaid, unleashing the divine power of an unlimited cornucopia.

TAKE ALMOST ANY ANIMAL in the world, multiply its metabolic rate by its life expectancy, and you come up with an interesting result: Most creatures spend about 1 billion heartbeats on this earth, no matter what size they are or where they live. Humans, perhaps because of advances in medicine in the past couple hundred years, are a bit above average; we can expect 2 or 3 billion. Hummingbirds probably clock out between 1 and 2 billion.

If this seems too simple to be true, consider a quick calculation.

Assuming that a human's heart rate is 70 beats per minute at rest and 60 beats per minute while asleep, that we sleep roughly a third of our lives, and that we live about seventy years, we can expect our hearts to beat 2.45 billion times.

Hummingbirds are trickier, because their heart rate is so variable. Let's assume that a typical hummer's heart rate is 1,200 beats per minute in flight, 250 beats per minute at rest, and 50 beats per minute while asleep. It might be flying a quarter of the time it is awake, and might sleep through a third of its seven-year life. In that case, a hummingbird's heart would beat 1.26 billion times.

The rule applies to mice, elephants, and most animals in between, and works because larger, longer-lived creatures tend to have slower metabolisms. It doesn't account for disease or sudden death before old age, and it's obviously a generalization. But it does indicate that the hearts of most animals are wired to put in about the same effort, no matter how fast they run.

People with lower resting heart rates generally live longer, which fits right in with the billion-heartbeat rule. This points to one of the many benefits of exercise. Going for a run might speed up your heart for a short while, compressing more beats into a small amount of time, but it will also lower your resting heart rate as you get fitter, ultimately extending your life.

Slow down your baseline, in other words, and you'll live longer.

Psychologists have demonstrated that the pace of life in large cities—how quickly post office transactions are processed, how much time a stranger will spend answering your question, how accurately bank clocks are set—is correlated with the speed of pedestrians on the sidewalk. In 1999, a psychology professor at California State University analyzed walking speed in cities from thirty-one different countries, and found that the pace of life was fastest in Japan and western European nations, and slowest in economically undeveloped countries. People moved more quickly in colder climates and in "individualistic cultures." Faster cities had significantly higher rates of smoking and heart disease. Switzerland was fastest. Mexico was slowest.

In 2007, a separate study found that the differences in walking speed among countries could be dramatic. Residents of Singapore averaged 10.5 seconds to cover 60 feet of sidewalk, while people in Bahrain took closer to 18 seconds to meander the same distance, and Malawi was off the scale at 31 seconds.

The same study also compared its measurements with data collected ten years earlier, and came up with a somewhat shocking result: In the previous decade, walking pace had increased by 10 percent overall. In the world's largest cities, people were physically moving faster, averaging 3.27 miles per hour compared with 2.97 miles per hour just ten years earlier.

Are we becoming more like hummingbirds?

Hummers are so fragile that they leave few fossils, so our knowledge of their origins is punctuated by long gaps. We do know that hummingbird-like creatures existed in what is now Germany about 30 million years ago, that they must have evolved most of their incredible adaptations for high-powered flight since then, and that they are most closely related to swifts and nightjars. As they gradually grew smaller and faster, their engines more refined, their equipment more specialized, hummingbirds became the miniature turbos that we admire today.

But their flight in the fast lane came at a high cost. If they ever pull over for a rest, they die of starvation. Their legs, shriveled up to save weight, are too weak to take even one step. Hummingbirds are slaves to speed, desperately fighting for control of calories, so single-minded that they don't even partner up to raise a family. They apparently have an unusually high rate of heart attacks and ruptures, which is hardly surprising. Hummers blast through their billion heartbeats in one brilliantly intense rush, and when the engine shorts out, they fade just as quickly into aether, hardly leaving any trace to show that they ever existed at all.

It seems like humans are speeding up—we strive for more gratification with fewer delays. Our fast-food culture isn't a cliché; it's a fact. And things are only accelerating. But do we really want to become hummingbirds?

The famous golfer Walter Hagen, perhaps the first athlete ever to earn a million dollars, recognized the need to slow down once in a while. He might even have been pondering hummingbirds when he once quipped: "Don't hurry. Don't worry. And be sure to smell the flowers along the way."

part two

MIND

fight or flight

WHAT PENGUINS ARE AFRAID OF

The first time I walked through the Adélie penguin colony at Cape Crozier, Antarctica, I quickly learned to step carefully. Researchers have scrutinized aerial photos and counted 300,000 penguins here—about the population of Pittsburgh—crammed into a single valley about one mile wide, next to the edge of the Ross Ice Shelf in a remote, frozen region claimed optimistically by New Zealand. The birds nest within pecking distance of one another, in swirling, dense swatches that, in satellite images, resemble a grainy photocopy of a Rorschach test. On the ground, it's chaos. I found myself sidestepping along the edges of two-foot-tall crowds within the metropolis, detouring around stubborn penguins that wouldn't get out of my way.

It was 20 degrees below zero Fahrenheit with a biting wind, and I wore a full polar kit: insulated white bunny boots, black snow pants, layers of long underwear that would not be changed for weeks at a time, rainbow-tinted ski goggles, and a red down jacket with its hood lined in coyote fur. The penguins regarded me like I'd just dropped in from outer space.

Personally, I'd be a bit frightened if a gigantic, multicolored alien suddenly appeared next to my house, but these birds weren't afraid, despite having had relatively little human contact. They were simply curious, often approaching with timid steps if I paused quietly for a moment. Penguins untied my shoelaces, hesitatingly preened the sides of my sleek pants, and fell in line behind me in a drawn-out game of follow-the-leader. The followers stopped when I stopped, walked when I walked, and froze like a group of guilty thieves whenever I suddenly turned on them. As I wound my way deeper into the valley, fifty penguins followed in a gaggle.

Such trust is remarkable in a wild creature; perhaps it helps explain the star appeal of penguins—why a bird in one of the world's most remote places is also one of the most well-studied species on earth, and why, as the actor Joe Moore has been credited with observing, "it is practically impossible to look at a penguin and feel angry." The arrival of humans hasn't diminished the birds' energetic personality. They are still as approachable today as they were when the first explorers stepped onto the Antarctic continent.

"They are extraordinarily like children, these little people of the Antarctic world," wrote Apsley Cherry-Garrard, a twenty-five-year-old adventurer who visited Cape Crozier in 1911 during Robert Scott's ill-fated South Pole expedition. "Their little bodies are so full of curiosity," he observed, "that they have little room for fear."

But even penguins have their limits, as I soon discovered. As part of a long-running research project called Penguin Science, I had been deployed to Cape Crozier with two other researchers to spend three months in a rough field camp—no shower, laundry, fresh food, or resupply. One of our most important goals was to track penguins with several biotelemetry global-positioning-system tags, recently designed, which could not only transmit their location in real time but also measure temperature, pressure, and light levels underwater. That meant that with a bit of luck, we would be able to visualize how the birds hunt for fish. First, though, someone had to go wrangle some penguins.

I usually tell people that catching a penguin is easy: You just walk over to one and pick it up. But this glosses over a certain wiliness on the part of the penguin. Yes, penguins are extremely approachable, especially when on a nest; they will let you walk within two or three feet of them without much concern. Invade that bubble of personal space, though, and you'd better be quick. If you grab the feet with one hand, scooping under the belly, and hold the base of the tail in the other, you can have the penguin contained in your lap before it realizes what's happening.

Attaching a global-positioning-system tag takes just long enough to apply several strips of adhesive to the bird's back. When I released my first patient, it blinked for a moment, shook itself off, then waddled placidly back to its nest, settling down to a state of quiet meditation with hardly a glance in my direction even though I still crouched just a few feet away. I wondered what it could possibly be thinking. Did it even connect me with what had just happened to it?

I wrangled several penguins in this way to apply the tags, and it was the same story each time: immediate calm upon release. They didn't hold grudges. The birds were just as easy to grab a few days later when I returned to remove the tags. They maintained their good-natured curiosity toward me despite my interference.

It's slightly harder to get hold of a penguin that isn't on a nest. On one occasion, I found a bird meandering aimlessly with a metal flipper band (used to track individuals from one year to the next) that was bent out of shape. To fix the band, I'd have to catch the penguin. This required backup, so I radioed one of the other researchers, who brought a long-handled net.

The wandering penguin was as wily as could be. It would let me approach within ten feet with complete unconcern, but not one step closer. When I invaded that bubble of personal space, it backed away as quickly as I advanced. The bird was fast, too; when I sprinted at full speed across the ice, it easily ducked my grasp, sidestepping like an expert matador as I slid past. Only by flanking the penguin between us could the other researcher and I get close enough to try to snag it with the net. Even so, it took several tries and a final, heroic dive to pin the bird down.

I could tell the penguin was panicking as we disentangled it from the net and, in just a few seconds, fixed up its band. Like others I'd handled, it bit my armpit and bashed my chest with its flippers—a surprisingly strong force from an eight-pound bird barely taller than a bowling pin. When I let it go, it danced off a few steps until it had regained its ten-foot buffer zone, sighed, and immediately lost all visible interest in me, as if I didn't exist.

FEAR IS AN EMOTION shaped by danger, so it's no surprise that birds in remote regions are generally approachable; they've lost their fear of people—or, more likely, never learned it in the first place. This is especially true in polar areas and on isolated islands where humans have ventured only recently. The very remoteness of the existence of these birds protects them.

One of the best places to experience this effect is the Galápagos Islands, where visitors often comment that the local wildlife has no fear. Camera-toting tourists delight in walking right up to boobies, albatrosses, iguanas, and sea lions without causing the least fuss; it's enchanting to be so intimately close with wild animals.

Even Charles Darwin was taken by the tame nature of Galápagos wildlife. When he visited the islands in 1835, the young naturalist tried an experiment on a marine iguana that was basking on a sunny beach. Darwin walked up to the three-foot-long iguana, picked it up by its tail, and, without warning, flung the reptile as far as he could into the ocean. (Darwin didn't show much compassion for Galápagos iguanas. He often killed and ate them, referred to them as "stupid," and nicknamed them "imps of darkness.") In response, the iguana swam to shore and crawled right back to the spot where it had been basking, still at the naturalist's feet. Darwin leaned over, picked it up again, and again threw it into the ocean. The iguana crawled back to its spot, looking confused. This cycle was repeated many times, even though, as Darwin later pointed out, the lizard could easily have escaped by swimming lazily away along the shoreline.

The iguana-tossing experiment fascinated Darwin. Fourteen years before he penned *On the Origin of Species* and became history's most famous naturalist, he published a book called *The Voyage of the Beagle* describing his five-year trip around the world. Of twenty-three adventurous chapters, only one was dedicated to the Galápagos, but Darwin found plenty of space to ponder the "apparent stupidity" of the tossed iguana.

"This reptile has no enemy whatsoever on shore, whereas at sea it must often fall a prey to the numerous sharks," he mused. "Urged by a fixed and hereditary instinct that the shore is its place of safety, whatever the emergency may be, it there takes refuge." In other words, iguanas aren't completely fearless; they are just more afraid of sharks than anything on land.

Galápagos birds are likewise nonchalant about people, but, like Antarctic penguins, only up to a point. The Galápagos National Park regulations do not allow tourists to approach wildlife beyond a threshold of disturbance—if the animal reacts to you, you're too close. In the case of nesting seabirds, this might be as near as three feet. Try to nudge closer than that, and you'll get a daggerlike beak aimed at your face. The rules are there to protect people as well as birds.

This disturbance threshold is called flight distance—the

distance at which an animal begins to react in the face of approaching danger—and has been measured to study fear in all kinds of animals. Of course, fear of humans doesn't always translate to other fears, and flight distance varies with context. But because it's easy to quantify, flight distance is often used to indicate perceived levels of risk.

Flight distance generally decreases with the disappearance of humans and other large predators, which makes sense: There is little benefit to being fearful for animals in predator-free environments. Experience defines the best way to react. In the cases of Antarctic penguins and Galápagos iguanas, this dictates indifference; running away just wastes energy if there is no danger of being eaten. But in the rest of the world, most wildlife—besides a few animals with special defense mechanisms, such as rattlesnakes and tigers—is much more skittish.

Interestingly, flight distance can also be shortened at the opposite end of the spectrum. Urban animals that are continuously exposed to humans, like house sparrows and European starlings in city parks, have become conditioned to us. That's why urban pigeons will walk right under your feet: They've learned that humans aren't a threat and might even give them some food. It's also why scarecrows, plastic owls, hawk silhouettes on windows, flapping flags, noisemakers, and other bird scare tactics generally don't work very well. Over time, birds can become conditioned to almost anything as long as they remain safe.

So it's the animals in the middle—the ones with some experience of humans, but not too much—that are most afraid of us.

WHEN THE SEA ICE at Cape Crozier broke in midsummer, the penguins must have been happy. For months, they had been commuting miles on foot across the frozen ocean between their nests on land and any crack where they could slip under the ice to hunt fish in the shadowy world below. An overnight windstorm fractured the ice and blew it offshore, leaving large expanses of open water near the penguin colony. The penguins no longer had to walk very far to go hunting.

For the first time, I could watch them dive into the sea, and if I stood with my toes at the edge of the ice and peered straight down, I could even see the penguins swimming underneath my feet through the crystalline water. Below the surface, they compensated for their clumsiness on land. Penguin wings are called flippers for good reason—they help the birds slice through the depths like two-toned torpedoes. I marveled at the long trails of bubbles as each penguin cut and swerved after fish.

The birds were eager to take advantage of their sudden access to the ocean and lined up at the edge of the ice like kids on a diving board. Instead of jumping straight in, though, they would dawdle interminably on the brink. At first, I thought they were being polite, each waiting their turn; they usually formed an orderly bunch with the earliest arrivals at the front.

Then I realized that they were afraid of the water.

Several penguins would line up at the edge and stare downward as if they were searching for something. More would waddle up from behind in straggling groups until dozens of them were standing in a tight, nervous bunch. No penguin wanted to be the first to get its feet wet. The birds in front would scoot sideways and work their way to the back of the group as more piled in from the rear. Eventually, so many of them would be jostling for position, with the ones in the back pressing forward, that some poor penguin at the front would get shoved off the edge. A second after it flopped into the water, an invisible stoplight would suddenly flip green: The entire group, sometimes a hundred penguins or more, would dive in all at once.

Cherry-Garrard noticed the same behavior in 1911. "They will refuse to dive off an ice-foot until they have persuaded one of their companions to take the first jump," he wrote, before suggesting that the off-key singing of one of his shipmates, who liked to serenade the penguins with a song called "God Save," would "always send them headlong into the water."

This dance was funny to watch, but the penguins had good reason for it. Ten-foot-long, 800-pound leopard seals—the top predator in Antarctica, besides killer whales—sometimes prowl the beach along Cape Crozier in hopes of ambushing an unsuspecting penguin. The birds were wise to be cautious.

If penguins have nightmares, leopard seals probably play a starring role. The seals, recognizable by their large, gunmetal-gray body with a faintly spotted pattern, elongated snout, and devilish grin, have knifelike incisors designed for tearing apart fish, penguins, and other seals. Leopard seals usually can't catch a penguin on land or in the open ocean, but they have learned that the birds are vulnerable in the shallows as they enter and exit the water. So the seals wait, submerged, like killer submarines just offshore, for an unlucky penguin to jump into their jaws.

Even people are occasionally victims of leopard seals. In 2003, a twenty-eight-year-old scientist working for the British Antarctic Survey was snorkeling near the Rothera research station, on the Antarctic Peninsula, when a leopard seal suddenly appeared, grabbed her, and dragged her sixty feet underwater in front of a group of horrified colleagues. By the time a rescue boat could retrieve her, she was dead.

A member of Ernest Shackleton's Imperial Trans-Antarctic Expedition in the early 1900s was chased by a leopard seal until his companions shot it. And more recently, researchers using inflatable boats have had to add reinforcements to prevent punctures because seals have been known to chomp on the pontoons. On a continent where the total human population usually does not exceed 5,000, these incidents seem to occur with somewhat alarming frequency.

Imagine, then, what must pass through a penguin's mind as it contemplates diving into the water. No wonder it doesn't want to make the first move. Penguins might not be afraid of us, but when confronted by a cold, dark ocean with invisible lurking danger, they conceivably feel as much fear as we would in the same situation. They have to deal with this threat every time they get in the water, every time they search for food.

This kind of penguin fear has lately caught the attention of scientists. The danger of leopard seals, some suggest, hasn't just affected the way penguins line up on an ice floe; it might also clarify much larger, more complex aspects of penguin behavior. For instance: Why are penguins afraid of the dark?

WHEN AN ANIMAL IDENTIFIES serious danger, its system gets ready to cope—the classic fight-or-flight response. The heart beats faster, the lungs process more air, energy floods into the muscles, bowels give way, and digestion slows. All of these physical reactions help a creature stand to battle, like a buffalo facing off against a wolf, or sprint to safety, like a gazelle running from a lion.

Some of the physical effects of fear aren't so easy to explain. Shaking, loss of hearing, paralysis, and even fainting can follow a sudden fright, and the wisdom of passing out in front of a hungry predator is debatable. The fight-or-flight mechanism has been criticized for being too simplistic: Many animals have other reactions when faced with danger, such as employing camouflage (though this could be seen as flight). Fear is not an on-and-off emotion; it encompasses a spectrum of reactions, many of which don't ignite an emergency physical response.

Researchers have even suggested that males of some species, including humans, evolved different strategies for dealing with fear than females. The fight-or-flight scenario may be male-oriented. Female animals of many species often confront danger by sheltering their offspring and seeking a group of companions, in what is now called the tend-and-befriend response. This difference in the reaction to fear could have evolved in humans when men and women were largely segregated in their work, and has even been theorized to account for the greater life expectancy of women, as the effects of fight-or-flight mentality may be more damaging to overall health. The different responses could also have resulted from more recent cultural conditioning; nobody really knows, but it's interesting that men and women may confront their fears in different ways.

Still, the fight-or-flight response is common among many animals, and the mechanism is straightforward. It is triggered by the release of stress hormones, including cortisol and epinephrine—also called adrenaline—which cause the physical manifestations of fear.

How the brain decides to release those hormones is complicated. Fear may be one of the most basic and widespread emotions. In the 1980s, the psychologist Robert Plutchik hypothesized an emotional wheel, like the color wheel, which has become influential. Plutchik's wheel contains eight primary emotions arranged in paired opposites: joy and sadness, trust and distrust, surprise and anticipation, and anger and fear. Like colors, emotions can exist in varying intensity—fear, for instance, ranges from minor (unease) to extreme (horror). By combining those basic building blocks, Plutchik argued, any secondary emotion could theoretically be formed: for example, submission is fear combined with trust and awe is fear combined with surprise.

Plutchik also believed, along with many others in his field, that emotions evolved because they increase survival—and that animals experience emotions in the same way we do, which is particularly important. The first of ten emotional postulates in his book *Emotion* states, "The concept of emotion is applicable to all evolutionary levels and applies to all animals as well as humans." This means that penguins may be emotionally similar to us. If that's true, those of us trying to get inside the mind of a penguin must first understand our own.

Although emotion itself is innate, fear of particular things is regulated by experience. Humans and other animals can learn to fear just about anything under the right conditions, and most conscious fears seem to be learned. The association of specific events with anticipated consequences is called fear conditioning, and experiments have shown that negative reinforcement is a powerful force. In order for a stimulus to result in a response in any animal, all that matters is that the two occurrences coincide repeatedly so that the brain learns to associate them. The most classic example may be Pavlov's dog; the Russian physiologist rang a bell every time he put food in his dog's tray and proved that the dog soon salivated at the mere sound of the bell.

In extreme cases, fears can be linked to seemingly random events. A well-known example is Little Albert, a nine-month-old infant studied by American psychologist John Watson, along with his assistant Rosalie, in 1920. According to Watson, Albert was a normal, healthy child with an aversion to loud noises and no inherent fear of white lab rats. In Watson's experiment, he introduced a white rat to Albert, who was allowed to happily play with it for a while. Then Watson changed tactics: Whenever Albert reached to touch the rat, Rosalie clanged a hammer into a steel bar just behind the baby's head. Each time this happened, the infant burst into tears at the sudden noise, and, after several repetitions, he would cry at the sight of the white rat alone. Watson reported that Albert then began to cry in reaction to other things resembling a white rat: a Santa Claus mask with white cotton balls, a furry dog, and a white coat. In a very short time, Albert had learned to be afraid of something benign.

The Little Albert experiment is considered unethical by today's standards, and it didn't turn out well for anyone involved. Watson had a scandalous affair with his assistant, Rosalie, divorced his wife, and was kicked out of his university. He never got the opportunity to recondition Albert, who presumably remained terrified of fluffy white objects. But the study inspired decades of research about fear, and John Watson turned the course of psychology toward behaviorist ideas that favored nurture over nature.

Conditioning is mostly a good thing—it allows our brains to know how to respond to our surroundings without analyzing every detail. Even fear conditioning is usually beneficial because it allows us to avoid danger. Some doctors have used the same concept to treat addicts with aversion therapy. The idea is that alcohol paired with, say, a vomit-inducing drug will teach an alcoholic to associate the two and then quit drinking. This technique can be dangerous because it adds more abuse than it eliminates and doesn't treat the underlying condition, but it sometimes works.

Fear has recently been linked with an almond-shaped part of the mammalian brain called the amygdala, nestled deep inside the skull, which is known to process memory and emotion.

The amygdala probably plays a role in guiding our fight-or-flight response—that feeling you get, for instance, within half a second of a close lightning strike. The reaction is immediate, innate, and, in a sense, unconscious. Fear also gets routed through the cognitive parts of the brain, which take a few seconds to catch up—eventually directing you to run for cover. By the time rational thoughts take over, your body is primed and ready to make an escape. The two pathways—a subconscious, immediate response from the amygdala and a delayed, logical response after the rest of the brain realizes what's going on—have been nicknamed the low road and the high road.

The separation of these two pathways was nicely illustrated by a brain-damaged woman who suffered amnesia in 1911. She was incapable of reasoning or forming new memories, so each time she visited her doctor, he would have to reintroduce himself and explain why she was in his office. One day, when she entered the room, the doctor shook her hand while concealing a sharp pin in his palm. He then left for fifteen minutes. When he returned, the woman predictably had no memory of who he was or how he had pricked her. But when the doctor reached to shake hands again, she wouldn't do it. Asked why, she couldn't explain; the woman merely said, "Isn't it allowed to withdraw one's hand?" Although her high-road logic had been damaged, her subconscious low-road response remained intact.

Birds don't have an amygdala, but researchers have hypothesized that other, similar structures in the avian brain might have evolved to perform a similar function. They certainly react to immediate threats in the same way we do: Frightened birds become agitated, vocalize, freeze in place, or try to fly away. Most examples of fear in birds are of this type of low-road response; once a bird escapes past a certain, ingrained flight distance, it loses all noticeable interest in a predator.

The question of a high-road response to fear in birds is harder to answer, and there is less evidence for it. Think about all the penguins that were clobbered for food by Antarctic explorers in the early 1900s. It was ridiculously easy for Scott and Shackleton to kill all the penguins they could eat—they merely strolled up to the unsuspecting birds and bashed them over the head (though they eventually learned that seals tasted much better, and left the poor penguins alone). Like Darwin's iguanas and other persecuted Galápagos wildlife, the rest of the penguins never seemed to catch on and remained as approachable as ever despite the repeated murder of their comrades. The birds evidently couldn't infer that the abuse befalling their friends might also happen to them. Humans in the same situation would probably have learned to avoid the new danger by using an analytic response.

This doesn't mean that fear in birds isn't learned behavior. Birds that are hunted regularly, such as waterfowl, are known to be more easily spooked, with greater flight distances, during the hunting season than they are in other months. Likewise, birds that are regularly exposed to predators are more wary than those that live in predator-free environments. They can learn fear from personal experience—and they can learn fear from their parents. One study of quail found that chicks born of fearful mothers but raised by calm foster parents grew up to be calm adults; in other words, their fearfulness was more affected by how they were raised than by their instincts (though the degree to which this was true was moderated by the genetics of their true parents).

Birds can also be conditioned to be less fearful. A study of robins on an island in New Zealand found that one generation after the eradication of rats from the island, the robins showed significantly less agitation when confronted with a model rat than did a population of robins on a nearby island where rats still roamed free. In just that one generation, they'd been conditioned to live in a predator-free environment.

But these examples still don't achieve the kinds of cerebral, high-road fears—such as long-term emotional stress, worry, and dread—that humans have. Can birds project into the future based on emotional memories? Do they have lingering, non-instinctual anxiety? This is still unanswered. There is an essential difference between anxiety and fear. Anxiety is a mood without an identifiable stressor, but fear is an immediate reaction to a threat; of the two, fear is much easier to study because its physical symptoms are obvious.

So in birds and in humans, fear itself is innate, but the timing of the response is often learned. When it comes to our instinctual reactions to danger, we are quite similar. But beyond that, it's tough to know what birds are thinking, whether they have long-term worries.

Psychologists Susan Suarez and Gordon Gallup, Jr., neatly summed up the situation in a volume of the scientific textbook *Bird Behavior.* They allowed that birds experience emotions, just as people do, but pointed out that we don't have a very good framework for them. "The concepts of fear and emotionality can only be meaningfully applied to birds," they said, "when they are explicitly tied to events which would have adaptive significance under natural conditions"—the assumption being that birds *do* have some kind of emotions, just like us. Penguins have feelings, too.

IN SPRING AND FALL, the sun rises and sets in Antarctica on a regular schedule. Days are brilliantly bright and nights are abysmally black.

In between, for a couple of months during summer, the sun refuses to go down; it roves around the sky in a tilted, counterclockwise circle, dipping close to the horizon in the wee hours but never quite touching. During my entire field season at Cape Crozier, I never saw one sunset, and never saw any stars. In winter, the opposite is true: Months pass without a single sunrise.

Penguins, like us, are tied to the sun. The birds are active during daylight, and they sleep at night. In winter, they migrate to more northern areas outside the zone of perpetual darkness.

But why?

It's not like penguins can't function around the clock. In the constant daylight of summer, penguins ditch their circadian schedule. They go on feeding trips that last for days. While I set an alarm to keep a twenty-four-hour rhythm, I noticed that the penguin colony still remained as active at midnight as it had been at noon, and that penguins took naps whenever they felt tired, snuggling down in a comfortable-looking flat spot against the ice at any hour. Forget New York—the real city that never sleeps is the penguin metropolis at Cape Crozier in midsummer.

When the sun begins to roll below the horizon in late summer, though, the penguins align their schedules with it. They rest at night, go to sea at dawn, and return by dusk. When winter arrives, they swim hundreds of miles north to escape the months of constant darkness.

For years, scientists assumed that, like humans, penguins can't see well in the dark. That would certainly hamper penguins' ability to eat during Antarctic winter, when the sun spends most of its time shining on the other half of the world. Fish may even be easier to catch at night because they have a harder time detecting their predators in darkness. If the penguins could navigate after the sun went down, they might have an easier time finding food.

But recent research has shown that penguins can see in darkness just fine. Our global-positioning-system tags with temperature, pressure, and light sensors indicated that the birds are often catching fish between 50 and 100 meters beneath the surface, a depth that is always as gloomy as early night. Sometimes they successfully forage even deeper. Emperor penguins occasionally swim as deep as 500 meters—about a third of a mile below the surface—where they might as well have their eyes shut. We wouldn't be able to function down there, but penguins apparently are able to pursue fish in near-total blackness.

So why do they avoid the dark? Perhaps it's just more convenient to be active when the sun is up. Hunting fish is one thing; socializing must be more fun in daylight.

The scientists who published the depth studies proposed a different explanation. They suggested that fear of leopard seals and killer whales drives penguins out of the water at night. Yes, it might be easier to catch fish in the dark, but penguins would be at greater risk of being caught themselves—and as we know, penguins are terribly afraid of leopard seals. They will commute miles over land instead of swimming the same distance parallel to shore, just to avoid the risk of those gnashing teeth.

An ecology of fear could explain the penguins' winter migration, too. Unlike most migratory birds, which commute between food-rich destinations, penguins travel to marginal feeding areas in the winter with *less* fish than the year-round, nutrient-rich waters surrounding the edge of the Antarctic continent. They could be avoiding bad weather or thick sea ice that blocks access to fishing grounds. But fear of predators lurking in the blackness could also drive penguins northward, where they would have some reassuring daylight during the darkest months.

It's an interesting theory, and it could easily be true. I witnessed the brutal terror of leopard seals one late summer afternoon at Cape Crozier. A seal had been prowling up and down the beach, just offshore from the penguin colony, for a couple of hours. Groups of hunting penguins would go skipping away in a panic whenever they swam too close to it. The attack happened suddenly. A lone penguin wandered down to the shore, failed to take proper precaution when entering the water, and waded in without spotting the danger. In a microsecond, the seal darted forward, snapped its jaws shut, and began to thrash the unlucky bird like a Rottweiler with a chew toy.

No wonder penguins go to great lengths to avoid being caught. Leopard seals are known to play with their prey before eating it, and this one took about 20 minutes to consume his meal in stages, using his sharp incisors to skin the penguin raw. Gore streamed through the seal's teeth; the wild-eyed, ugly gray head reared up like something from a horror movie, jaws snapping, blood saturating the freezing water, a chunk of the penguin's gleaming rib cage dangling from a corner of its mouth. As I looked on, mesmerized and horrified, I couldn't help thinking: If that bird had just been a little more careful, it would have escaped. Sometimes, even for penguins, a little fear can save your life.

beat generation

DANCING PARROTS AND
OUR STRANGE LOVE OF MUSIC

When Dr. Aniruddh Patel first watched the video of Snowball, as he later reported to *The New York Times*, his "jaw hit the floor." Patel is an associate professor of psychology at Tufts University. Snowball is a sulphur-crested cockatoo—a type of large, white parrot with an ornate yellow crest, native to Australia—who likes to dance. In 2007, destiny, in the form of YouTube, brought the scientist and the dancing bird together.

If not for the Internet, Patel might never have met Snowball, who became a viral sensation after his owner in Indiana posted some impromptu footage of the bird's dance moves. In the video, the cockatoo capers on the back of a chair to the Backstreet Boys song "Everybody (Backstreet's Back)." He sways rhythmically, bobs his head up and down, energetically waves his feet, and twists in time with the sound track, looking for all the world like a teenybopper rocking out to those repetitive chords and bass lines. The video of the dancing bird had 200,000 hits in one week, and, at last count (five years later), more than five million views on just that one clip.

Patel had never seen anything like it. All his life, he'd been fascinated by music and the brain. When his Ph.D. adviser, the luminary American biologist E. O. Wilson, sent him to Australia to study ants in 1990, Patel tried to emulate his professor's devotion to tiny insects. But one day he realized that he was more interested in the human biology of music. "You must follow your passion," Wilson reportedly advised, so Patel gave up studying ants, left Australia, and dedicated himself to a groundbreaking thesis in language and music at Harvard University.

That launched him into the field of neurobiological research, which was just then beginning to focus on brain imaging; Patel subsequently published, among other things, a fascinating paper showing that our brains process language and music in much the same way.

Snowball also bounced around for a bit before making his mark in the world. Like many pet parrots, he rotated through a series of owners. Nobody knows where he came from or who raised him as a chick. When he was about six years old, he was adopted by a family in Indiana, who kept him for several years. But when the daughter in the family went off to college, Snowball began getting cranky and aggressive. The girl's father decided that the parrot wasn't getting enough attention, so he relinquished Snowball to a rescue shelter called Bird Lovers Only, which specializes in caring for neglected birds, also handing over Snowball's favorite CD: the Backstreet Boys. "Just play it for him and watch what happens," the owner suggested.

Irena Schulz, the founder of Bird Lovers Only and a former molecular biologist, about died laughing the first time she witnessed Snowball in action. Lots of parrots had passed through her shelter, but none had quite the same charm as this one. She made a short video of the dancing parrot, uploaded it for fun, and, well, things snowballed from there.

The quirky cockatoo was soon featured on *CBS Sunday Morning*, *The Morning Show with Mike and Juliet*, *Ellen*, *The Tonight Show with Jay Leno*, *Late Show with David Letterman*, NPR, BBC, CNN, National Geographic News, Animal Planet, a commercial for a new Taco Bell beverage, a commercial for Loka bottled water in Sweden, and three TV shows in Japan. Fans mailed in CDs from all over the world, hoping Snowball might dance to them (a German polka caught his interest). Visitors started knocking on the door at Bird Lovers Only, purportedly to adopt a parrot, but really just to get a glimpse of the celebrity.

Schulz booked Snowball's guest appearances as if she were a Hollywood agent. She recognized Snowball's sudden fame as an opportunity to promote awareness for unwanted parrots, which often outlive their owners, and—why not?—to offer custom T-shirts, buttons, CDs, DVDs, and bumper stickers bearing the freshly trademarked Snowball name. An official Facebook page is now mostly an outlet for fanciful "Snowball: The Dancing Parrot" cartoons; the latest one, when I checked, featured Snowball competing in dressage at the Olympics. YouTube has its own channel where you can watch Snowball gyrating to everything from Queen to Lady Gaga.

But Schulz never dreamed that her rescue bird celebrity might be scientifically important. Ani Patel, who promptly contacted Irena Schulz when he first saw Snowball's video, also wasn't sure at first. The neurobiologist was stunned by the bird's apparent ability to dance to a beat, but wasn't convinced that it hadn't been trained, or that its owner wasn't leading its movements off camera. Could Snowball the dancing cockatoo truly synchronize his movements to music?

Patel and Schulz conducted a little experiment to find out. Patel used a computer program to alter Snowball's favorite Backstreet Boys song, "Everybody (Backstreet's Back)," into eleven different versions, each on the same pitch but with different tempos, from 20 percent slower to 20 percent faster than the original. Then Schulz played each version to Snowball on camera while she stood quietly in a corner, watching the bird perform by himself. At slower speeds, Snowball sometimes swayed from side to side, slow jamming, while in the speeded-up trials he spontaneously slammed his feet up and down in time to the tune. Mostly, he stuck to one basic move: a simple head bob.

Patel meticulously analyzed the videos of each trial, marking the exact frame when each of Snowball's head bobs reached its lowest point. Then he compared those time points with corresponding beats in the music to find out whether they coincided.

It wasn't a perfect match. Snowball often drifted ahead of or behind the beat, and in some trials, especially at slow speeds, refused to dance at all. Patel isolated instances of synchronized dancing among longer periods that were not matched to the tempo of the music, called them synchronized bouts, and found that such periods accounted for only about 25 percent of the time Snowball spent dancing—meaning that three-quarters of the time the parrot was well off beat. If anything, Snowball preferred to bob along at faster speeds; when all movements were combined, the bird's motions averaged quicker than the underlying tempo.

But Patel still thought that Snowball was moving rhythmically. The researcher took a closer look at the synchronized bouts he'd picked out, performing a statistical test to determine whether those periods could have occurred randomly. The result was highly significant: The probability of Snowball displaying even as much synchronization as he did merely by chance was minuscule.

That satisfied Patel. Along with a couple of coauthors, including Irena Schulz, he published a paper called "Experimental Evidence for Synchronization to a Musical Beat in a Nonhuman Animal" in the journal *Current Biology*. The paper itself was noteworthy. It marked the first time that any animal, besides us, had been shown to coordinate its movements to an external musical rhythm—a characteristic that had been regarded as uniquely human.

LOTS OF ANIMALS DANCE, in a broad sense. Under the general definition, "to move rhythmically," just about any creature can be said to dance at some point in its life. Dancing doesn't necessarily have to occur to a beat; even humans dance without music sometimes, and many animals make movements that might satisfy a more encompassing definition, for all kinds of unrelated reasons.

For instance, honeybees do a well-documented "waggle dance" to convey the location of food outside the hive; the angle of the dance refers to the direction and the distance to the food. Male rattlesnakes intertwine in a sinuous battle for the right to mate. The shovel-snouted lizard of Namibia dances across hot sand dunes to keep its feet from roasting. Cuttlefish habitually pulsate and turn colors to psyche out their prey. Short-tailed weasels perform a crazy "dance of death" that, as far as anyone can tell, may either serve to confuse prey or be caused by an infection in the brain. Spinner dolphins, when they leap wildly out of the water, are thought to be ridding themselves of parasites—or, just possibly, jumping for joy.

In wild birds, dancing is usually confined to seduction. Whooping cranes perform elaborate courtship displays complete with floating liftoffs, high kicks, and 360-degree ballerina turns. The bizarre New Guinea bird of paradise resembles nothing more than a shimmering flying saucer as he struts for the female, with specialized plumes dangling dark specks like a cloud of flies around his head. Clark's grebes take their dancing to the water, where, after a period of mutual head bobbing, they plane in pairs like jetboats across the surface of a lake, mimicking each other's movements à la synchronized swimmers.

Maybe the most accomplished avian dancers are the manakins, a group of related tropical bird species that are each smaller than a Rubik's Cube but display all the same bright colors. For manakins, being colorful isn't enough; females expect their mates to dance—and not one by one, but all at once. Male manakins gather in the dense, dark understories of South and Central American rainforests in the greatest dance contests of the bird world. Each species has perfected its own moves. The red-capped manakin, for instance, slides sideways along a branch in a passable rendition of Michael Jackson's moonwalk. The club-winged manakin claps its wings a hundred times a second—twice as fast as a hummingbird—to produce a mechanical trilling sound in short bursts. The most extreme species, such as the long-tailed manakin of Central America, use wingmen: Two males dance together in a tightly choreographed cartwheel in front of a prospective female, each vocalizing as they leapfrog over each other's backs. By previous agreement, one of them always gets to mate afterward while the other sits in the bushes nearby. A pair of male long-tailed manakins may work together like this for five years, building up their jungle reputation as hot dancers, before the alpha male dies and the backup dancer takes his place with a new apprentice. It's the only example of cooperative male-male displays ever discovered in the entire animal kingdom. Who says humans are the only species with boy bands?

But throw on a song like "Everybody (Backstreet's Back)" and even manakins would probably be at a loss. As intricate as their moves are, the choreography is always the same, to an unvarying tempo. And they dance only to their own vocalizations. Strictly speaking, this isn't dancing at all; most dictionaries mention both movement and an external beat in their first definition of the verb *dance*. Manakins have the moves, but not the rhythm.

None of these examples of dancing animals satisfies that narrower definition. Although their movements are interesting, they don't illuminate the evolution of dancing to a beat, which is a specific cognitive ability. That's what makes Snowball so special: He can adapt his dancing to whatever song happens to be playing, be it a slow jam or a rousing polka. His rhythm is a little rough—Patel compared Snowball's synchronization ability to that of a small child—but the parrot does get down when he hears music. Snowball is interesting because he has crossed a line that was once thought to separate people from all these other shimmying animals.

Which brings up the question: Is Snowball a weird one-off, or do other animals nurture as-yet-unrecognized dancing abilities? Perhaps South American manakins could collectively rock out to the worst of 1990s teen pop if someone would just load up the jungle with boom boxes. Because scientists have for so long assumed that animals can't dance, maybe they haven't looked hard enough for evidence. In a global search for nonhuman dancing talent, where would you start?

WHILE ANI PATEL was busy analyzing Snowball, a Ph.D. student of psychology at Harvard, Adena Schachner, was inspired to look for dancing animals in the same place Snowball got his break—on YouTube. Schachner was particularly interested in various theories of how music and dance first developed in humans. Psychologists can't seem to agree on even basic facts concerning the origins of music, and Schachner thought that animals might give some clues about our own evolution.

Casting a wide net, she decided to scientifically analyze as many YouTube videos as possible of dancing animals for indications of real rhythm. This meant trolling through page after page of search results for terms such as "dancing cat," "dancing bird," and "dancing monkey," and independently scoring each video on a set of guidelines: presence of an animal, rhythmic sound, periodic movement, and other general criteria to narrow the field. Most videos were clearly not candidates for her study, and Schachner quickly descended into the bottomless online abyss of cute puppies, funny cats, and talking hamsters.

One of the top results for "dancing dog," for instance, features a canine named Carrie, wearing a dress, performing a full merengue dance with her owner while balancing on hind legs. (Space does not permit full treatment of the video's comments, generated from nearly 15 million views, but the highest-rated comment was "I have reached the end of the Internet. Guess I'll turn around and just keep going.")

Alternatively, a search for "dancing cat" yields any number of ridiculous animations of pets set to various sound tracks, such as "The Kitty Cat Dance," a timeless clip described as "the heartwarming tale of a cat's insatiable lust for provocative dancing." Viewer responses range from "I watch this every day" to "brain.exe has stopped working."

Schachner later lamented that the amusement of YouTube lasted for only a couple of hours, after which viewing the videos became something of an endurance test. Nonetheless, she amassed a collection of about 5,000 video clips, from which she selected a more-manageable 400 that each showed some animal moving in the presence of rhythmic sound. For any particular clip that suggested coordination with a beat, she used the same careful analytic techniques that Patel used with Snowball to determine whether the animals were displaying true rhythm. Common subjects were ferrets, dogs, parrots, horses, cats, albatrosses, pigeons, elephants, squirrels, dolphins, and fish, with dozens of others, including chimpanzees and orangutans.

In the end, only thirty-three videos showed signs of synchronized dancing. Parrots—of fourteen different species—accounted for twenty-nine of them; the other four showed Asian elephants.

At the very least, the results indicated that Snowball was not unique. Frostie, an up-and-comer parrot from California, has lately surpassed Snowball's YouTube view count with a lively interpretation of an aptly named song, "Shake Your Tailfeather." In the video's description, Frostie's owner declares, "Frostie can dance the socks off any bird on this planet!"

But besides parrots and elephants, no other animals made the cut. Despite analyzing dozens of dancing dogs and cats, some of which (especially dogs) had been trained for years to perform with their owners in elaborate competitions, Schachner could find no evidence that any of them were spontaneously moving to a beat—not even Carrie, the merengue star. Furry pets relied on their owners for direction or cavorted randomly to a musical backdrop.

The lack of evidence for primates is especially compelling. What could possibly set parrots and elephants—and us—apart from our close relatives, monkeys and apes?

Schachner nursed a personal theory, put forward earlier by Patel and strengthened by the study of Snowball, that the secret of dance is tied to an ability to mimic vocal sounds. Language, music, and dance are known to be tightly interrelated; excellence in one discipline may translate to the others. Patel had argued that only animals that are vocal mimics, learning to communicate by copying one another, have the potential to synchronize their movements with an external beat. In vocal-mimicking animals, cognition of sound is necessarily tied to motor skills. They hear something and move accordingly.

Relatively few animals are true vocal mimics, as far as we know: songbirds, parrots, hummingbirds, whales, dolphins, porpoises, walruses, seals, sea lions, elephants, some bats, and humans. The list includes a few creatures you might not expect, and covers both parrots and elephants—but, tellingly, no primates other than humans.

Brain structure seems to back up this theory. Birds that copy sounds have been shown to have modified basal ganglia, in parallel to the part of the human brain that helps identify musical beats. This could be the mechanical reason for an association of mimicry and rhythm: Parts of the brain that control auditory perception and motor skills have become physically more closely connected in animals that are vocal mimics.

The results of her YouTube study convinced Schachner that vocal mimicking helps predict the ability to perceive a musical beat. She also performed a separate case analysis of Snowball along with another famous parrot, a veteran African gray parrot named Alex that had been well studied by animal psychologists for decades (the bird's very name was an acronym for "avian language experiment"); neither Snowball nor Alex had been explicitly trained to dance, but both showed clear signs of spontaneous synchronization to music they'd never heard before. Finally, Schachner tried the same methods with a group of nine cotton-top tamarins, a type of monkey native to Colombia that weighs less than a pound. The tamarins may have been amused by Schachner's music, but they definitely didn't dance to it.

If all vocal-mimicking animals were able to dance, though,

you'd expect hummingbirds and bats to join in, along with other species on the list. Why only parrots, elephants, and humans?

Both Patel and Schachner admit that it's not a perfect theory, going only so far as to say that mimicry is a precondition of dancing to a beat. They agree that other preconditions probably exist, perhaps including a social nature (which would immediately rule out most hummingbirds, for starters, as discussed in the chapter titled "Hummingbird Wars"), a superior cognitive standard (which would make sense in the context of parrots, elephants, and humans), and the desire to mimic movements as well as sounds. Snowball's owners have often danced with him, and he seems to copy their arm-flapping motions (though Irena Schulz insists that the parrot has invented his own moves and sometimes dances inside an empty room if music is left playing). Without a role model, Snowball may never have learned the behavior at all. The same could potentially be said for any pet parrot dancing on YouTube.

And even if the vocal mimicry hypothesis is true, it's unclear whether the idea has much greater meaning in the animal kingdom. Parrots in the wild don't dance, even though they are still excellent vocal mimics. Elephants in the wild don't dance, either, though elephants do show advanced learning and memory. We now know that parrots have the ability, but we can only assume that dancing to a musical beat has no practical benefit for creatures outside of human civilization. Dancing, in terms of wild animal behavior, is thus reduced to an intriguing psychological footnote.

For people, though, that assumption is less applicable. Some researchers believe that dancing parrots may bear on one of the greatest debates of human musical history.

WHEN IT COMES to the evolution of modern music, researchers are divided into two camps. Some say music is biologically useless—that it is a mere by-product of our large, complex brains. Others declare that music must have evolved through natural selection because it gave humans an adaptive benefit. The answer is shrouded in ancient history, but that doesn't stop people from arguing about it.

Darwin himself had a difficult time with the question. "Neither the enjoyment nor the capacity of producing musical notes are faculties of the least direct use to man in reference to his ordinary habits of life," he wrote in *The Descent of Man*, published in 1871. Yet, he observed, music is present "in men of all races, even the most savage."

All cultures have embraced music in some form. Because it is so universal, music must fulfill some part of being human. Musical ability is hardwired in a fundamental way; children begin to sing and dance without much guidance, and recent research has indicated that even sleeping newborns, two or three days old, can sense the beat in a series of drum rhythms.

But as Steven Pinker argued in his 1997 bestseller *How the Mind Works*, even though music helps define who we are, it doesn't necessarily help us survive. Pinker agreed with Darwin's assertion that music does not serve any practical use to humans; without it, we'd still be able to secure food, shelter, mates, and other basics of life. Other animals don't write operas or download tracks from iTunes, and they do just fine. Maybe it's for the best that vultures don't lose themselves in melodic contemplation.

Pinker suggested that the evolution of music in particular, and the arts in general, was simply a by-product of language and other complex brain functions, asserting that music is "auditory cheesecake"—designed to tickle the "pleasure circuits of the brain" like so many unnecessary fats and oils. While interesting to us, said Pinker, music is inherently superfluous.

As you might imagine, the whole auditory-cheesecake premise didn't go over well with artists, musicians, and historians, who believe that music deeply enriches our lives. And many evolutionary biologists also disagreed, suggesting that music could have evolved as an adaptive benefit of natural selection instead of as a by-product. This opposing camp offered a number of possible alternatives for the origins of music.

One popular theory states that our musical roots can be traced back to "motherese," the whispered communication between a mother and her baby that strengthens the familial bond. Modern parents coo to their children just as they probably did thousands of years ago; the way we change our voices in the presence of babies seems instinctual. Perhaps that instinct led us gradually to adapt our sounds to other situations.

Another idea is that music originated in seduction. Many animals, especially birds, sing beautiful melodies to prospective mates (even if they don't keep a beat). Our own songs could be an elaborate ritual of attraction that, like a peacock's tail feathers, was heavily selected for across many generations. Darwin concluded that this was probably the case, although he seemed unsure of himself; for inspiration, he looked at male gibbons, which sing to hold territories and attract females.

Darwin also mentioned another theory, not incompatible with any previously mentioned, that music predates language. He suggested that our imitation of earthly sounds as a simple means of communication coalesced into words, grammar, and modern syntax. (Noam Chomsky, the prominent American linguist and activist, suggests that grammar is universal across all languages, an indication that we all follow the same rules, however different we may sound.)

Other researchers have lately picked up this thread again. Psychologist Steven Brown has proposed the musilanguage hypothesis, which holds that at one point in our evolutionary history, language and music were one and the same. As Ani Patel found with brain imaging, we compute the two in similar ways, and language and music have so much in common that, in a sense, they are still indistinguishable. Both contain elements of pitch, tone, phrasing, melody, and rhythm. Either one can be used to communicate facts or convey emotion. In some languages, such as Chinese, changes in tone alone can differentiate words with separate meanings.

It's easy to see how we might have ended up with musical languages. Early humans, trying to convey meaning to their friends, began to imitate sounds from the world around them: wind, water, other animals. That imitation led to the formation of words, which gradually became more complex and abstract. Children learned to mimic the sounds that their parents made, and systems of communication passed down through generations.

What about rhythm? Does it relate to the basic human walking gait, as some suggest, or could it have evolved as a way to synchronize social groups—even as a method of putting armies into a battle trance? It seems likely that our ability to maintain a beat developed along with early music, but because sound leaves no fossils or other long-term traces, we may never know which came first, the melody or the beat.

Snowball's ability to synchronize to music shows a hidden talent in parrots that mirrors our own. If, as Ani Patel and Adena Schachner believe, rhythm arises from vocal mimicry, it could be evidence for the musilanguage theory. And in that case, the entire musical field would derive from imitation—a fine irony in today's copyright-obsessed music industry.

But results of Patel and Schachner's research also could go the other way, directly supporting Steven Pinker's assertion that music is just a quirky by-product of evolution. Schachner seems to think so, anyway. "If an observed behavior does not exist in the natural behavioral repertoire," she has written, "it has no potential to increase or decrease fitness and thus cannot be directly selected for or against." In other words, behaviors that don't exist in the wild, such as dancing to a beat, have no effect on an animal's ultimate survival, so they can't have evolved as an adaptive benefit. Parrots, at least, have no use for music except in the company of humans; their ability to sync to it shows that they must have evolved their ability as a by-product.

This line of reasoning could naturally be extended to us. If vocal mimicry incidentally caused parrots to be able to jam to the Backstreet Boys, why not people? The same mechanism could apply to both.

The idea is not as dispiriting as it sounds. Regardless of how we came to appreciate music, our enjoyment of it is no less meaningful. Many aspects of our modern culture aren't immediately linked to the rough-and-ready world of natural selection, and that's a good thing: It's what makes us human. Whether music is auditory cheesecake or a prehistoric battle cry doesn't change its powerful influence.

Just ask Snowball. His whole existence has been validated, at least in human terms, by the love of music. But please, for the sake of all that's feathery, someone teach that bird some musical taste.

seeing red

WHEN THE PECKING ORDER BREAKS DOWN

Let's talk chicken.

By the end of the twentieth century, domestic chickens outnumbered humans by about four to one on this planet, distinguishing them as the most abundant bird species on earth. Actually, chickens are the world's most numerous reptile, amphibian, mammal, or bird, period. At any given time, the globe hosts about 20 billion domestic chickens, though most don't live very long. The average North American eats more than fifty pounds of chicken a year (the equivalent of about twenty-seven individual birds), which ranks, pound for pound, slightly less than beef and a bit more than pork.

Birdwatchers tend to denigrate chickens because domesticated species don't qualify on official life lists. Unless you happen to prowl the tiger-infested jungles of India, where a few wild red junglefowl—the tropical ancestor of today's McChicken—still flock together, you won't get much credit for observing poultry. But we should pay attention to chickens, if for no other reason than their familiarity.

We can learn a lot about the world from farm fowl, as a six-year-old Norwegian boy demonstrated a hundred years ago.

THAT YOUNG BOY, tending his mother's chicken coop outside of Oslo, noticed something curious about the birds he fed every morning. When any two hungry chickens met at the food tray, one would always make way for the other, patiently waiting its turn. Instead of fighting like unruly teenagers over Thanksgiving dinner, the chickens usually formed an orderly line with minimal fuss.

Furthermore, the order was utterly predictable. One particular hen was always the first to eat, followed by a second individual, then a third, and so on. At the water dish, their behavior was the same. If one tried to jump the line, it was barraged by pecks from the birds in front, and it quickly retreated.

By the time he was ten, Thorleif Schjelderup-Ebbe was keeping his observations in detailed notebooks. He'd discovered that the order of the feed line was based on aggression; certain hens, for some reason, always dominated others. So Schjelderup-Ebbe—let's call him Thor—began charting aggressive interactions among the birds in his mother's chicken coop, hoping to figure out, scientifically, whether his ideas made sense.

When Thor tallied up his observations, a pattern emerged. The top bird, at various times, had pecked every single one of the other chickens in the coop, but had never been pecked in return. In second place was another hen that had pecked everyone else except the top chicken, and, accordingly, had been pecked only by that bird. This trend continued down the line until only one poor hen was left standing—barely—which had been pecked by every other chicken in the coop but never delivered a single peck of its own. The alpha chicken always ate first, and the lowliest one always got the crumbs.

It took Thor many years to accumulate these observations because his flock seemed so comfortable with their order that they rarely acted aggressively toward one another. Each bird knew its place. The lower-ranked hens accepted their status and preferred not to challenge up; altogether, it was a fairly peaceful if unequal arrangement.

The system wasn't always perfectly linear, though. In a few cases, A pecked B, B pecked C, but C pecked A. Thor called these situations "triangles," and, like the romantic plot of some Hollywood films, triangles were intriguing. Rather than trying to offer theories or explanations—others would do that, later—he continued to keep detailed notes about chicken behavior, eventually using the data for his dissertation in 1922. Thor termed his system of hierarchies *Hackordnung*, a German word that translates to "pecking order" in English. It was the first time anyone had used the term, which would come to be routinely applied to humans in everyday language by the 1950s.

This concept of social hierarchy wasn't really news for poultry growers. Chickens have been domesticated for roughly 4,000 years, and any old keeper could tell you that some birds dominate others in their coop. Over the years, farmers have learned a few basic truths about pecking orders in chickens. Dominance is related to size, but the correlation is not very strong; experienced and wily birds are much more likely to rise through the ranks than mere big brutes. And one sure way to incite fights is to introduce a new hen to an existing flock. Do it on a Friday, wise farmers say, so you can keep an eye on things for a couple days. Sneak new birds into the henhouse after dark when chickens are sleepy and new birds can't accidentally steal someone's favorite roosting spot. Don't introduce one new bird into an existing flock, but bring in at least two at a time so they can partner up and form an alliance.

But Thor was the first to put science behind the concept of social inequality in chickens. Today, his notebooks are referenced as a pivotal point by practically every academician who studies dominance hierarchies; some say he pioneered an entirely new field of behavioral research.

As with many geniuses of their time, Thor was not applauded for his efforts, which then seemed dangerously anthropomorphic. After a rival spread rumors that Thor had written an anonymous story criticizing his professor—who happened to be the first female professor in Norway—in the student paper,

Thor's reputation was irreparably damaged, and he never received his doctorate degree.

The man who coined the term "pecking order" was, ironically, pecked into submissive oblivion.

WE'LL COME BACK TO CHICKENS. But first we have to talk about tennis for a moment.

Every year in November, the world's top eight male tennis players convene for a high-stakes tournament called the World Tour Finals, an end-of-the-year championship that traditionally caps each season's grueling ten-month playing schedule. Since 1970, the event has rotated between fifteen different cities and it's been played on carpet, grass, and hard courts—both indoors and outdoors. It's been won by the greatest players of all time (Roger Federer, Pete Sampras) and one or two guys you've probably never heard of (Manuel Orantes, Michael Stich). Through it all, an unusual format has endured at the tournament.

Most tennis tournaments are played with classic single-elimination draws in which if you lose one match, you're out. Half of the players advance in each round until only one is left standing. Single elimination is straightforward, brutal, and occasionally unpredictable; if someone has one bad day, that's it. The format is also efficient at picking winners. In a field of 128 players, one needs to survive only seven matches to win the tournament.

But the World Tour Finals, with just eight players in the draw, is a different sort of event. Rarely do the top tennis stars get the opportunity to play just each other, and everybody relishes these high-profile matchups. Tournament organizers settled on a partial round-robin format instead of the usual single-elimination strategy. The field is divided into two groups of four. Within those groups, everyone plays everyone else, then the top two from each group advance to the semifinals, which sets up a traditional final.

In a sport like tennis, with relatively stable rankings and the purest talent at the top, the round-robin format seems to permit fewer flukes. If a good player has a bad day, he gets a second chance. A quick glance at the list of World Tour Finals champions from the past forty years confirms this idea; Roger Federer has won the event six times, Pete Sampras and Ivan Lendl five times each.

These round-robins can also get confusing, though. In 2006, three players in one of the groups—Andy Roddick, Ivan Ljubičić, and David Nalbandian—had each won one match, with a somewhat circular result: Roddick beat Ljubičić, who beat Nalbandian, who beat Roddick. The three-way tie was broken by advancing the player who had won the most sets (Nalbandian), but the denouement was unsatisfying.

Though round-robin tournaments may appear to result in true rankings, the reality is that they often don't.

Things did not go smoothly in 2007, for example, when the men's professional tennis organization decided to try the round-robin format at several other events. At a Las Vegas tournament that year, American defending champion James Blake was playing Argentinian Juan Martín del Potro for a slot in the quarterfinals when del Potro, who wasn't feeling well, called it quits halfway through the second set. Blake had built a commanding lead, and, had he finished the match, would have advanced easily. But because the match was stopped, he didn't qualify for the quarterfinals. Outcry was immediate and intense. At one point, it was decided that Blake would advance anyway; a day later, the decision was reversed. It was all very embarrassing and confused.

Within a month, the round-robin idea was quietly abandoned, and all regular-season tournaments reverted to single-elimination format. The experiment, though a failure for men's tennis, did draw attention to a longstanding question in the upper echelon of professional sports: Is there an inherent, absolute skill ladder among the top contenders that holds up under random matchups, or can only a brutal system, like single-elimination draws, yield a decisive list of rankings? The answer, it seems, lies somewhere in the middle.

What does any of this have to do with chickens? A lot, actually. Tennis rankings are a pecking order among athletes based on head-to-head matches. This system—from the World Tour Finals to your backyard chicken coop—is more triangular than you might think.

AN IDEAL PECKING ORDER is perfectly hierarchical, like the rungs of a ladder, with no two subjects of equal rank. This is what's known mathematically as transitivity: A is greater than B, which is greater than C, and so on down the list. Many things appear to be transitive—pecking orders in chicken coops, professional tennis rankings, even the body weights of ten random people lined up from largest to smallest. But some characteristics lend themselves more to this kind of hierarchy than others. Absolute values, like weight, will always be perfectly transitive. Other things we often think of as absolute—dominance in chicken coops and on tennis courts—aren't always perfectly measurable.

Some ornithologists have used transitive relationships as a

test of overall intelligence: Can birds infer rankings by applying logic to separate pieces of information? In the most straightforward example, a bird might be repeatedly given a choice between options A and B, and always rewarded for picking A and punished for picking B. Once it has learned to prefer A over B, the bird may be offered a new choice between B and a third option, C, but this time be rewarded for picking B and punished for picking C. In this way, after learning to prefer A over B, the bird will also learn to choose B over C. Then, the interesting part: What would that trained bird do with a novel choice, A vs. C? Will it pick A?

Experiments show that pigeons (and other animals, including monkeys) will generally choose option A, apparently making the mental leap that if A > B and B > C, then A > C. In practice, chickens could use this kind of reasoning to efficiently determine their pecking order. If one bird in the coop dominates another, and that one dominates a third, then the first one wouldn't have to prove itself against the third. But it's probably not that easy; this lab test may be too simple to explain what's really going on. For instance, the trained bird may pick A over C just because it has always been rewarded for picking A and punished for picking C in other pairings.

So the experiment gets more complicated. This time, the bird is trained with five options instead of three, to teach it that A > B, B > C, C > D, and D > E. Once it reliably learns each of these pairings, the bird is given a choice between B and D, each of which was rewarded half the time in training. Pigeons will, in fact, pick B over D, suggesting that they can track a transitive order. But this experiment may also be too simple. The pigeons might pick B just because they've often seen it with the dominant A, and they might reject D because it's often alongside the lowly E. This is the "value transfer" hypothesis, and you've seen it in practice when regular guys join the entourage of rich, Ferrari-driving friends to attract women—they're hoping women will prefer someone in the passenger seat of a Ferrari over some random guy on the street. Value transfer is real and measurable, but unfortunately it's practically impossible to devise an experiment complicated enough to rule out the effect—even with seven options, which is the most that a pigeon can usually be trained to separate in its lifetime.

The trouble with these lab tests is that, for a pigeon or a chicken or any other animal, pecking at colored buttons for a superficial reward has no inherent value—the birds have to be painstakingly trained to prefer one button over another strictly for experimental purposes. The pecking order in a chicken coop, on the other hand, has immediate, real rewards and consequences; the birds must quickly learn how to pick their fights or they'll get beat up.

Real-world evidence does suggest that chickens can use logic to infer their place in the pecking order. When chickens observe an unfamiliar bird facing off with a higher-ranking individual, they are later more likely to pick a fight with the newcomer if it lost. If the new bird beats the existing dominant chicken, others in the coop will usually leave it unchallenged. Nobody has been able to prove that these fight-or-not decisions affect the pecking order itself, so it's debatable whether chickens use implied logic to construct their hierarchy, but this observation is interesting.

If chickens can really weigh dominance relationships and fill in gaps without having to fight every other bird in the coop, then pecking orders can be seen as an expression of intelligence in a creature not usually known for its mental capacity. That would be pretty cool.

TRY THIS PEN-AND-PAPER EXERCISE to see for yourself how pecking orders work, often imperfectly.

1. Draw a large octagon on a sheet of paper.
2. Draw a line from every point of the octagon to every other point.
3. Think of any eight foods, and label each outside corner with one of them. Your diagram should look like a stop sign inscribed with an eight-pointed star shape, each corner labeled with a food.
4. Without thinking too hard, decide which food you like best from each possible pairing. Label each straight line with an arrow pointing toward your preference.
5. Count up how many "wins" each of your eight foods received.

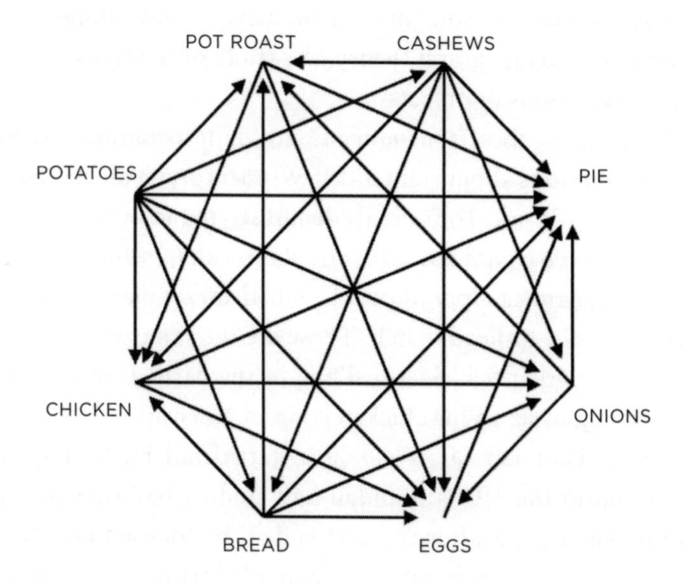

When I tried this exercise with pot roast, cashews, pie, onions, eggs, bread, chicken, and potatoes, I vastly preferred pie (seven votes) to potatoes (zero votes). No surprise there. Cashews and bread had one and two votes, respectively, and eggs, with six votes, ranked second only to pie—I'm a big fan of eggs. But the fascinating part of my food web was in the middle range. Onions, chicken, and pot roast had four votes each. A dead tie.

How could this be? I had envisioned a clear, descending scale of taste preference: The favorite (pie) would get seven votes, the least favorite (potatoes) would get zero, and the others would sort themselves out in between. If I had simply ranked the eight foods, it would have been easy to score them from zero to seven with no ties. But a different pattern emerged from the round-robin-style "tournament." I like pot roast better than onions, onions better than chicken, and chicken better than pot roast, but this leaves me with a triangle of onions, chicken, and pot roast. A > B and B > C, but, illogically, C > A.

This is exactly how mathematicians depict round-robin tournaments using graph theory, the study of relationships between collections of objects.

Triangles in these tournaments are quite common, despite our assumptions about rigid order. Winners aren't always clear, as the tennis World Tour Finals demonstrated in 2006. Mathematicians even have a name for circular results, calling a round-robin tournament "paradoxical" when every player loses at least one head-to-head match. These are the same triangles noticed by Thorleif Schjelderup-Ebbe in the early 1900s within the pecking order in his chicken coop in Norway.

Where Thor left off, sociologist/statistician H. G. Landau picked up in the 1950s. Landau had studied ballistics during World War II, but after the war ended, he focused his attention on biological phenomena, eventually settling on chickens.

Like Thor, Landau observed that, in any given pair of hens, one would always dominate the other, even if the two birds were separated and reunited years later. Landau analyzed in precise detail the hierarchies of chicken coops. Unlike his Norwegian predecessor, though, he focused more on the circular rather than linear nature of these relationships. As in our favorite-foods exercise, Landau assigned each bird a dominance score; he called the top-scoring birds "king chickens" and tried to quantify their relationships to the other birds.

Not only were there frequent triangles, but some coops had multiple kings, at odds with our general perception of pecking orders. Landau showed that it was even possible for every single chicken in a coop to be a king while they all maintained stable and unequal dominance relationships with each other, in one perfectly circular hierarchy. If pecking orders were Thor's legacy, then the idea that they could be circular became Landau's claim to fame. Backed by the mathematician's careful proofs, this concept has been handed down today in the form of Landau's so-called king chicken theorems.

These theorems, which quantify the possible results of graphical tournaments, apply to many things besides chickens. They show why those round-robin events often don't work in professional sports. The winner of a tennis match depends largely on how the two players' styles complement each other as well as other factors beyond skill, such as the court surface, weather, and audience. The best players constantly beat each other in one event and take revenge at the next. That's where great rivalries come from, and why we are enthralled by sports. Even the best players can lose to lesser opponents. And, as Landau showed, sometimes there is no "best" player at all, even if every individual matchup is perfectly predictable.

This is also why we have trouble ranking our favorite foods.

Theoretically, we should be able to arrange a strict, hierarchical list, but the differences between adjacent entries on that list are often small enough that other variables—presentation, mood, perception—have a greater impact when we are deciding what to have for dinner. Like a tennis match, favorite foods depend on immediate conditions as much as absolute flavor.

Pecking orders break down completely in chicken coops when the total number of birds exceeds about thirty, probably because chickens have trouble recognizing more than thirty familiar faces. At that point, the birds can't keep track of one another closely enough to enforce dominance, and, instead of a transitive society, the group becomes egalitarian. All members are socially equal. (The numerical limit for social groups entails its own fascinating field of study; check out Dunbar's number, which holds that typical humans are limited to about 150 friends in their social sphere, beyond which acquaintances are essentially strangers.)

Egalitarian societies sound great, but they have their own problems. In large chicken coops, such as industrial farms with tens of thousands of birds crammed together, aggression is still present—but, with no established order of dominance through which to channel those feelings, aggressive actions are dealt out randomly. Sometimes this even leads to bloodlust, an ugly manifestation of chicken personality that seems senseless in today's high-productivity farm operations. If we could completely weed out aggression from chicken society, wouldn't we all be better off?

One man thought so. In 1989, he set out to bring peace to the world, one chicken at a time.

EVEN WHILE RANDALL WISE attended college and founded a successful software company in Boston, he couldn't stop thinking about chickens.

Randall's father, Irvin Wise, had managed a chicken farm in Northern California since the 1960s, and the old man always took an intense interest in chicken behavior, often involving his son in serious discussions about poultry. Mostly, why did those darned birds fight so much? If the hens could get along better, their farm would save money, because stressed-out chickens ate more feed, expended less energy on laying eggs, and nursed unnecessary pecking wounds. Randall listened carefully and, working with his father, hatched an idea.

In recent years, chickens have been bred to grow ever faster and bigger. Today's six-week-old chicken is six times heavier than an equivalent breed in 1957 and has about 10 percent more breast meat. But breeders haven't focused much on personality; as chickens have grown in size, they've also become more aggressive. After all, chickens were first domesticated for cockfighting, not food.

Randall noted that for some reason the color red caused his chickens to go berserk. This fact was already well documented by other chicken farmers, who occasionally watched in horror as their birds pecked each other to death. When a chicken bleeds, other birds in the coop become fascinated by the bright red wound and peck at it repeatedly, sometimes inflicting serious damage.

Chickens have incredibly well-defined color vision, even better than ours. Unlike, say, bulls—which are color-blind and chase the matador's cape because it's twitching, not because it's red—chickens nurse a real lust for blood. The color red incites them to violence.

Other farmers had tried replacing the regular lightbulbs in their chicken coops with red ones, which mitigated some of the chickens' bad behavior. They theorized that the red light caused red objects—combs, wattles, blood—to blend in better, making it harder for the birds to single them out. But the darker light was difficult for humans to work in, so most farms switched back to regular illumination and trimmed their chickens' beaks with a hot knife, hoping to blunt them enough to prevent serious chicken-on-chicken injuries.

Randall pondered all this while he spent eight years cooped up in the software industry. Finally, he just couldn't stand it anymore and sold his business for millions, headed west, and founded a new company, called Animalens, to design red contact lenses for chickens.

That's right. Tiny red contacts. For chickens. Rose-colored glasses didn't work; he'd already tried it. The frames wouldn't stay on the chickens' round heads.

Randall believed he had found the perfect recipe for success. When the birds wore his contacts, they'd see their surroundings bathed in a soft red glow, as if the coops were wired with red lights. The birds would be calmer, more productive, and less apt to kill one another. Randall believed that the lenses could save poultry growers hundreds of millions of dollars per year through increased productivity and decreased feed consumption. Animalens contacts hit the market in 1989, priced at 20 cents per pair, or 15 cents apiece for large orders.

The idea wasn't as outlandish as it sounds, especially considering the weird breeds produced by the chicken industry. We've got the frizzle, a chicken with curly, mutated feathers; the phoenix, which grows tail streamers up to twenty feet long; the silkie, a fluffy cotton ball on legs; and the Transylvanian naked neck, sometimes called the turken for its wild resemblance to a turkey, with bare skin from the shoulders up. Why not a chicken that always sees red?

Initially, large-scale chicken operations were intrigued. Chickens wearing Randall's red contact lenses did seem to behave more calmly, saving an anticipated third of a penny for each egg produced. Considering that one Wisconsin-based chicken farm now produces 7.5 million eggs *per day*, the cost savings would be substantial for high-volume egg plants.

But farmers hated to hold each bird's head steady while they tried to jab contacts into millions of little eyeballs, and installing the tiny pieces of plastic was harder than Randall had let on. The lenses were supposed to stay in for life, but they kept falling out or, worse, causing physical damage. Chickens wearing contacts developed severe eye irritations and actually became more stressed. One investigation showed that birds wearing contacts developed bigger heart muscles, perhaps to cope with the anxiety of having irritated eyes and no fingers with which to soothe them.

Ultimately, the red contacts did little to reduce operating costs, and animal-rights groups were mad as wet hens about their effects on the birds' health. By the mid-1990s, the company had folded.

Randall Wise thus learned a worthy lesson: Messing with basic social instinct is a risky business. Even if pecking orders are confusing to us, for chickens they are a matter of life and death. Aggression and dominance may not always be desirable, but they do serve a purpose. Trying to erase those characteristics might just cause more problems than it solves—in a chicken farm and in the world in general. Sometimes, it's better to let a natural order sort itself out.

May the best chicken win.

cache memory

HOW NUTCRACKERS HOARD INFORMATION

Things were not going well for thirty-five-year-old scouting agent William Clark (of Lewis and Clark fame) on August 22, 1805. He was exploring the rugged Salmon River canyon, in what is now northern Idaho, on a typically harrowing day: A hunter from his group had just returned from a close run-in with some Native Americans, the terrain was brutally steep and rugged, and he didn't know where he was going. But sometime that afternoon the explorer noticed a bird interesting enough to record in his now famous journal:

"I saw to day [a] bird of the woodpecker kind which fed on Pine burs it's Bill and tale white the wings black every other part of a light brown, and about the size of a robin."

He didn't have time to elaborate (or punctuate) further, but his partner, Meriwether Lewis, described the same bird less than a year later with better detail and leisure, while waiting patiently for snow to melt in the Bitterroot Mountains on the expedition's eastward return journey. Lewis correctly surmised that Clark's black-and-white bird wasn't a woodpecker but a corvid—related to crows and jays.

"Since my arrival here I have killed several birds of the corvus genus," Lewis wrote. "It is about the size and somewhat the form of a Jaybird . . . it resides in the Rocky Mountains at all seasons of the year, and in many parts is the only bird to be found."

Lewis's specimens of this bird were eventually deposited in the Peale Museum in Philadelphia. Alexander Wilson included an illustration of the new bird in his 1811 book *American Ornithology*, calling it Clark's crow, with proper credit to the discoverer. (To be fair, Lewis was honored with his own namesake, the Lewis's woodpecker, which he had described in his journal the day before writing about the new crow-like species.) The name stuck but was later modified, and we are left today with an explorer's legacy: the Clark's nutcracker.

If you've spent time in the rugged mountains of western North America, you've probably heard and seen nutcrackers. The birds are easy to identify: big, black and white overall, with flashing white wings and tail patches in flight and a cool gray body. No other bird looks anything like a nutcracker. And they make themselves obvious, perching on exposed treetops and flying around high meadows, searching boldly for their next meal. Particularly fearless nutcrackers are sometimes called "camp robbers" for their habit of stealing food from unwary backpackers. Even if you hiked blindfolded into the mountains, you'd still know the nutcrackers were there. The characteristic loud *kkrrraaaaaaack* provides a running sound track in high-elevation pine forests, where the noise echoes off craggy peaks just as it did hundreds of years ago.

Lewis and Clark gave a fairly accurate description of nutcrackers. Lewis observed that they remain in the high mountains "at all seasons of the year," even while most other birds migrate south or to lower elevations for the winter. These days, one of the best ways to see a Clark's nutcracker is to visit a ski resort during peak season, where the birds often hang around high-elevation parking lots looking for handouts. But those early pioneers probably didn't realize that nutcrackers also, improbably, *nest* in the winter, laying their eggs during the brutal snowstorms of late January and early February. While the Lewis and Clark expedition wintered by the muddy mouth of the Columbia River, waiting for summer to arrive so they could head home, the nutcrackers they'd observed in Idaho were happily raising chicks high up in the mountains. By the time the snow had melted enough for the expedition to traverse east in May and June, the young nutcrackers had already fledged.

Not many other birds nest on mountaintops in the middle of winter, and for good reason. It's cold and stormy up there. Food is also scarce in the high country during the winter, and what little food may be left is difficult to find under packed snow. That's the main reason most mountain birds spend their winters in warmer climates—they'd starve if they didn't migrate. But nutcrackers are smart. They're related to the intelligent crows and ravens, and they've learned a trick to not only survive but thrive in the mountains during the cold season. The trick is pretty simple: They do what Lewis and Clark did.

Those great explorers, when Thomas Jefferson charged them to find a route across North America, knew that they were in for an unpredictable adventure, so they loaded up on supplies. Lewis reportedly spent $2,324 on gear—equivalent to about $50,000 today—for his thirty-man crew. The packing list included, among many other things, twenty-five hatchets, forty-five flannel shirts, five hundred rifle flints, thirty-five oars, an iron corn mill, eleven hundred doses of emetic (to induce vomiting in case of poison), twenty pounds of glass beads, a four-volume dictionary, three bushels of salt, and 193 pounds of "portable soup." One of the largest sections of the list was titled "presents for Indian tribes encountered." Every expedition marches on its stomach, and Lewis made sure that the Corps of Discovery would be well supplied to traverse the wilderness. He figured they could trade trinkets with the locals when hunting food became difficult.

Nutcrackers do the same thing to survive cold winters at high elevations: They stock up. The birds have a specialized diet of pine seeds, which are easy to find when cones ripen in summer and fall. But the cone crop is seasonal, so Clark's nutcrackers must cache massive quantities of pine seeds to eat during the winter and spring, when cones are unavailable. Amazingly, they depend completely on these stored seeds throughout the nesting season. Nothing else is available during the stark, white winter. Storing food makes it possible for nutcrackers to stay in the mountains, and even raise their young, throughout one of the harshest times of the year.

Because nutcrackers have to store a lot of pine seeds to survive an entire winter, during the bountiful part of the year they are forced to become veritable cache machines. The birds begin working full-time when cones first ripen in July and continue to gather seeds throughout the late summer, fall, and early winter. To help them carry food, nutcrackers have developed a special pouch under their tongue that can hold up to about a hundred seeds at a time. They use their heavy bills to pry apart pinecones and load up on seeds until the pouch is full, then fly off to stash their harvest in the ground—sometimes several miles away from the source tree. Instead of hiding them all in the same place, nutcrackers divvy up their loot in lots of little caches, poking three or four seeds at a time into the soil.

And this is where nutcrackers transcend mere survival. In one fall season, a single nutcracker may store tens of thousands of pine seeds in as many as 5,000 different mini-caches. The birds don't mark the spots, and there is no surface indication that something is buried below ground—in fact, the caches are often covered by snow when winter arrives. Nutcrackers may leave caches alone for nine months before returning to collect their contents, and they don't use the same locations from one year to the next. Sometimes other critters dig up the seeds, and sometimes they spoil underground, so the birds generally cache more than they need; one side effect is that extra seeds sprout, which helps the pine trees spread. Incredibly, hungry nutcrackers are able to locate most of their stashes through the winter.

It's a radical mental feat: Nutcrackers somehow remember exactly where thousands of different clumps of seeds are buried without a single yellow sticky note, global-positioning-system waypoint, or silly mnemonic.

How do they do it?

WHEN NELSON DELLIS, a twenty-eight-year-old former software developer from Miami, attempted to summit Mount Everest in 2011, he took along a pack of playing cards. Not for entertainment, exactly. Dellis had recently become involved in the cultish sport of memory games, and he wanted his brain to stay active while he climbed. During his ascent, he regularly sat down with the cards, carefully shuffled them, and spent a couple of minutes trying to memorize the order of the entire deck. According to habit, he kept track of how many seconds it took to flawlessly recall the fifty-two cards in sequence each time, and he noticed something odd: As he gained elevation, his recall times shrank. The concentration exercise was easier than expected in distracting and exhausting conditions, which encouraged Dellis. It was good to have a sharp mind; he was climbing to raise money for Parkinson's research, after all, and, while chasing high-elevation dreams, he was also training for a different kind of challenge.

Dellis was almost as happy about his memory times as his near-summit of Everest (he turned around 280 feet from the top). In the rarified pantheon of elite "mental athletes," he had become an instant star a couple of months earlier by winning the 2011 USA Memory Championship, a niche competition for speed memorization. The contest had been founded some fourteen years earlier by a former IBM executive to demonstrate the capabilities of the human brain, but then people like Dellis started signing up—and turned simple memory games into a sport in some ways as grueling as mountain climbing.

Trust human beings to find a way to compete over something as ordinary as remembering a list of names. One of the events at the championship—which is held every year in a meeting room at the corporate headquarters of Con Edison, the main electricity supplier in New York City—involves quickly memorizing 117 different photographs of unfamiliar faces, each image inscribed with a person's name. Competitors must also recite a fifty-line unpublished poem after only fifteen minutes of study and absorb as many random digits as possible in five minutes. Then there is Dellis's specialty, memorizing the order of a shuffled deck of cards.

In practice, Dellis has correctly recalled fifty-two cards in order after only thirty-three seconds to glance at the deck. At the competition, though, his best time is sixty-three seconds. Distractions and nerves play havoc with one's ability to concentrate—hence Dellis's card tricks at twenty thousand feet. In an effort to showcase drama at the USA Memory Championship, finalists must perform mental feats while sitting onstage, facing an audience (yes, there are spectators, commentators, and TV cameras), though the event still feels a bit like a room full of students taking the SAT.

Make no mistake, memory competitions are serious. There is a worldwide rankings list. The World Memory Championships, held each December in a rotating list of international cities, offers tens of thousands of dollars in prize money and is usually dominated by European and Asian competitors. In the United States, mental sports traditionally have held less prestige, but that is beginning to change. In 2012, journalist Joshua Foer popularized memory contests with his bestseller *Moonwalking with Einstein*, describing how he had trained to win the USA Memory Championship in 2006. After Nelson Dellis won it, he was featured in *The New Yorker*, on CNN, and in *Forbes*. Dellis now works full-time as a "memory consultant," gives inspirational speeches about climbing and memory, and spends hours exercising his brain.

When I last checked the world rankings list, the top slot was taken by Johannes Mallow, a German fellow who had swept first place in a streak of memory competitions in Sweden, Germany, and England, achieving the honored title of Grand Master. Mallow's official stats are impressive. In five minutes, he has accurately memorized 500 numerical digits and 85 random words. After just one hour of intense study, he has been able to perfectly recall 2,245 digits and 1,144 shuffled playing cards (twenty-two decks!) in correct sequence. When asked to memorize years of fictitious events, Mallow absorbed 132 dates in five minutes.

If this all sounds like something from *Rain Man*, it's not; these people do not call themselves savants, and they insist they have average memories—if they're not paying attention, they forget their car keys just like anyone else. Photographic memory, at least in the popular sense of being able to instantaneously absorb pages of phone books, is a myth. Nobody in the world can regurgitate reams of random information with just a glimpse (if anyone could, surely they'd show up to collect their $10,000 prize at the World Championships). Competitors train hard, like other athletes, and have developed complex systems to pull off seemingly inhuman feats of memory.

How do they do it? Do human memory athletes use the same techniques as Clark's nutcrackers?

IN HOPES OF LEARNING how birds memorize things, a graduate student at Northern Arizona University, Stephen Vander Wall, focused his attention in the 1970s on the remarkable ability of Clark's nutcrackers to remember where their snacks are buried. Vander Wall figured that nutcrackers might relocate stashed seeds in one of five ways: (1) The birds dig randomly and find buried seeds by chance; (2) They search randomly, but only in certain, heavily cached areas; (3) They are able to smell buried seeds; (4) They mark the ground surface in some way to indicate where caches are hidden; or (5) They remember the exact locations as if plotted on a mental map. With these five possibilities in mind, Vander Wall set out to design an experiment to whittle down the list.

He could see that following the birds in their natural habitat—across some of the most rugged country in the American West—would amount to a wild nutcracker chase, and that controlled laboratory tests would give much clearer results. So Vander Wall conducted his experiments with captive birds. He covered the floor of a large aviary with a couple inches of loose soil, and then arranged various perches, rocks, logs, and other "landmarks" around the room. After two individual nutcrackers, code-named Orange and Red, had been trained to bury seeds inside the aviary, Vander Wall began his tests.

First, he released Orange and Red inside the aviary and watched in separate trials to see where they made their caches. When both birds had stashed at least 150 seeds in the loose soil, in alternate sessions so they couldn't watch each other, Vander Wall snuck in and buried 100 seeds himself. He also removed 50 of the birds' caches. Then he let Orange and Red back into the aviary to see which seeds they would dig up. If the birds foraged by remembering the locations of their own caches, Vander Wall reasoned, they wouldn't be able to find each other's stashes or his additional buried seeds.

And that's exactly what happened. Orange found 63 of his own seeds, but none of Red's or Vander Wall's. Red found 61 of his seeds, three of Orange's, and none of Vander Wall's. The birds were not searching randomly, or they would have found an equal share of each group of seeds. The first two hypotheses—two variants on random searching—could thus be neatly ruled out.

Both birds also dug unsuccessfully for their personal caches that had been secretly removed by Vander Wall, which indicated that his third hypothesis—the birds found food by smell—was also false. That left just the fourth and fifth options: Either the birds had somehow marked the soil surface or they were able to remember locations in relation to nearby landmarks.

Vander Wall conducted a second experiment to investigate whether the birds might be marking the soil. He let Orange and Red cache additional seeds inside the aviary, then raked half of its floor smooth to obscure any ground cues. If they were looking for surface disturbances, then Vander Wall predicted that the birds would be able to find their caches only in the unraked half. But when the two nutcrackers were returned to their aviary, they proceeded to dig up seeds in both the raked and unraked areas, indicating that surface texture didn't impede their ability to remember locations. Hypothesis number four was rejected, too.

Only one possibility remained, that nutcrackers were using spatial memory to locate their stashes. To test it, Vander Wall designed a third, elegant experiment. As in the first two tests, he kept Orange and Red inside the aviary until they had made numerous seed caches in the loose soil. Then, as before, he removed the birds and made some secret modifications. This time, he rearranged the landmarks in half of the room, moving each of the rocks, logs, and other perches exactly twenty centimeters in the same direction; in the other half, he left the landmarks alone. He predicted that the nutcrackers would be able to locate their caches only where the nearby landmarks had been left in place.

When Orange and Red returned to the aviary after a period of fasting, they searched for their stashed food. This time, they found only half of the seeds—the caches in the undisturbed half, just as Vander Wall had predicted. In the other half of the aviary, the hungry birds dug about twenty centimeters away from the actual locations of their caches, almost exactly corresponding to the distance and direction that the local landmarks had been moved. The only exceptions were a few caches in the middle of the aviary, between the two zones, for which the birds dug at an offset of about ten centimeters.

Vander Wall showed that Clark's nutcrackers draw on spatial memory to remember locations—a pretty impressive feat, considering how many thousands of points they memorize every year. The basic idea is intuitive. The birds must build a three-dimensional mental map, plotting the locations of temporary caches within their visualization. They can remember where they store food only by tying each individual spot to an already intimate knowledge of landmarks within their home territory. This method orders the information in space, which turns relatively abstract knowledge into a useful foraging map that helps them get through the winter.

It's as if the birds are telling themselves: *Dinner is on the stove, my car keys are on the bedside table, and I left the car parked in the space by the blue lamppost.*

ACCORDING TO LEGEND, the Greek poet Simonides (556–468 B.C.), a charmingly witty, miserly international man of lyricism—dubiously credited, among other things, with inventing four letters of the Greek alphabet—was once invited to a banquet to celebrate the victory of a boxer. At some point during the festivities, Simonides rose from the table and stepped outdoors for a moment of solace; during that brief interval, the building collapsed and killed everyone inside, leaving the poet with a narrow escape. Excavation revealed the bodies of his friends to be unrecognizably mangled, but Simonides realized that he could remember where everyone had been sitting just by closing his eyes and visualizing himself at the banquet in the moments before he stepped out. Thus, he was able to identify the corpses.

This feat supposedly led him to invent a new way of memorizing information, a necessary skill for poets past and present. Simonides imagined himself walking through a familiar setting—say, inside a palace—and, as he mentally traced his path, he would paint vivid images at various points along the way. If he wanted to remember a poem that started with a line about a lion, he'd imagine a lion sitting on the front steps of the palace. If the next line talked about the moon, he'd think of the moon crammed into the entryway. And if the next line dealt with a beautiful woman, Simonides would picture her waiting at the foot of the staircase after he'd squeezed past the moon. He found that by using this strategy, he could keep many images ordered in his mind, and retain the nuggets he needed to remember.

At least, that's the story. Simonides' accomplishments are now chronicled only on a few papyrus fragments, so nobody knows whether he really escaped from a collapsing building 2,500 years ago (although it's clear that his poetry and ideas did affect the celebrated Classical period of Greece, and by extension all of Western civilization). But the memory-palace method of memorization—often called the method of loci by today's psychologists—endures, and continues to be used by those who need to absorb large amounts of ordered information, including memory athletes such as Nelson Dellis and Joshua Foer.

You don't have to imagine a palace. Just pick any familiar space—your childhood home, your commute to work, the interior of your favorite restaurant—and picture moving through it as you normally would. The key is to tag vivid images throughout your journey at points where you'll remember to look for them when you mentally retrace your steps.

At the USA Memory Championship, all of the top contenders use memory palaces to absorb packs of playing cards and lists of random digits and words. Individual approaches vary, but the strategy is always the same: Hang that information on a familiar spatial frame. If you can convert a pack of cards into, say, a list of celebrities strewn throughout your house, the deck becomes a story—and a narrative of surprises is much easier to remember than a bunch of meaningless numbers and suits.

Nelson Dellis, who broke his own record and defended his championship title in 2012 by memorizing the order of 303 random numbers in five minutes flat, has insisted that he has no natural talent for memory. He hadn't even heard of mental athletes just three years before he won the event. Anyone, he says, can train to speedily cram information. Memorization stunts are all about forming associations and creating story lines from static data, even if, as in the case of numbers and playing cards, those stories have nothing to do with the content.

To prepare for the random-numbers event in the contest, Dellis painstakingly pre-memorized a list of 999 people, each one associated with a number, action, and object. Number 124, for instance, might be Tiger Woods hitting a golf ball. Number 423 could be George Bush driving a limousine, and 858 could be Britney Spears singing in her underwear. When confronted with a list of random digits, Dellis breaks the string of numbers into chunks and translates them into composite images of people, actions, and objects, in that order, so that 124-423-858 would naturally bring to mind Tiger Woods driving in his underwear. (858-124-423, by contrast, would become Britney Spears hitting a limousine, and 423-858-124 would represent George Bush singing to a golf ball—you get it.) The possible combinations are nearly endless, which keeps the images fresh. As he works through a list of numbers and their resulting pictures, Dellis concentrates on placing the images chronologically in one of several memory palaces from his everyday life, so that when the time comes to regurgitate the list of numbers, he can simply envision walking through a familiar setting with these crazy scenes happening at intervals, and recite the random digits in accurate order. It takes practice to make the associations quickly—Dellis works at it for up to six hours per day—but the technique is basic.

The same method applies to memorizing a pack of cards; each card has been previously associated with a specific person, action, and object so that they can be vividly coded into a memory palace. The more outlandish, bizarre, and surprising an image is, the more likely you will remember it. Don't just think of the TV set in your living room, in other words—imagine that it's on fire, and oozing miniature unicorns, with George Bush slapping his underwear on top. This is exactly how Nelson Dellis was able to recite the order of a shuffled deck after looking at it for only 63 seconds, and win the USA Memory Championship twice in a row. The power of association is strong in the human mind.

It may seem odd that we find it easier to absorb new information by *adding* to the load of data. Why is it more reliable to remember a number alongside Tiger Woods singing in a limousine than to simply recall the number on its own? Why do we have to construct elaborate stories merely to memorize the order of cards in a deck—which, by itself, is just fifty-two short entries? You'd think the additional layers of nonsense would require more gigabytes of storage space.

The brain-computer metaphor goes only so far, and in this case is misleading. Brains and hard drives accomplish some of the same memory functions, but they don't operate the same way; though they both store information, they access memory differently. On a computer, a file is stored at a strict location—if you don't know its address, you can't find it. Brains seem to behave more like a search engine, using what's known as content-addressable memory, meaning they bring up information by using subjects and keywords. Often, when searching for an obscure memory, the brain will find what it's looking for only after another thought sets it off—that familiar eureka feeling of having something on the tip of your tongue when it pops up in an unrelated conversation.

This search-engine method is generally a good system for organizing vast stores of information. In one sense, the brain is one huge database of knowledge—but, again, the analogy doesn't quite work. Various scientists have tried to estimate the capacity of their own brains in bytes, as if the brain were a computer's hard drive, which becomes a vaguely circular exercise. According to one line of logic, the human brain has about 100 billion neurons, each of which might be able to store a single piece of information, so one human brain could conceivably hold a couple terabytes of memory—about the same as a few of the latest laptops. Others point out that neurons aren't isolated; if each one could connect with 1,000 other neurons, we'd actually have about 2.5 petabytes of storage—on the same order of magnitude as the total data processed by Google every day. Still others estimate that a brain might be able to store multiple exabytes (1 followed by eighteen zeros) of data, somewhere in the range of the combined capacity of all digital storage devices on earth today. These wildly varying numbers at least clearly illustrate our lack of understanding of the brain's functions and the futility of comparing analog and digital, organic and robotic machines.

Still, there are a few intriguing similarities between the memories of computers and brains. Computers store information in a strictly defined hard drive, and brains also catalog long-term memories in one location, a region called the hippocampus. Many studies of memory have tried to relate hippocampus size with the ability to remember information, and there does seem to be a correlation. But again there are differences. While computer hard drives aren't used for processing, the hippocampus is integrated with the rest of the brain and helps perform active functions. It is heavily involved in imagination, and it is the primary region used in spatial navigation. So we shouldn't be surprised that, unlike a computer, we can remember images better than digits. Our hardware is set up to associate memories with vision and other senses.

Clark's nutcrackers and self-professed mental athletes share a trait that computers don't: While boasting uncanny memories, both birds and people primarily use spatial techniques to recall facts. Whether it's pine seeds or playing cards, the brain—any brain—needs a story—any story—to latch on to important information. Sometimes, a picture really is worth a thousand words.

Remember that spatial map that nutcrackers use to track down cached pine seeds? It's a memory palace. Perhaps the ancient Greek poet Simonides should have looked to the birds for inspiration instead of a collapsing banquet hall. We struggle to believe that bird brains can remember where tens of thousands of seeds are stashed, but as it turns out, they use the same methods we do—which should give us hope for our own species.

We like to think the human brain is superior, but a recent experiment showed that birds are better at caching seeds than we are. A graduate student was pitted against a captive nutcracker, each burying dozens of pine seeds inside an aviary, then, after the passage of time, both digging up as many of their own caches as possible. The nutcracker beat the student by a large margin in a rare head-to-head cognitive win for birdkind. But in hindsight, it wasn't really a fair contest. The canny nutcracker had had a lifetime of practice, and from its standpoint, caching seeds was a life-and-death proposition; it would starve if it ever forgot where its food was hidden in the wild. The graduate student had no such practice or motivation. Winners of memory championships believe that our brain responds to exercise like a muscle; if we can teach ourselves to memorize packs of cards in seconds, we could also learn to memorize thousands of food caches (assuming we wanted to bury our groceries around the yard). Perhaps if someone like Nelson Dellis went up against a nutcracker, the result would have been different. Not even computer scientists can quantify the limits of our own memory—the mind boggles itself on the subject.

The brain's incredible mental capacity is granted with a clichéd caveat: Use it or lose it. In normal human brains, the hippocampus shrinks by one or two percent per year in adults (up to about five percent a year in Alzheimer's patients), and an idle hippocampus may shrink faster. In one fascinating study of chickadees, wild-caught birds lost a staggering 23 percent of their hippocampal volume just *five weeks* after being brought into captivity; caged birds, the scientists reckoned, had less necessity to navigate, interact, and remember information than their wild counterparts, so their brains shriveled (this measurement is known to fluctuate seasonally in the wild, too). But research has also indicated that this loss is negotiable, and that the brain may maintain itself better when regularly challenged. Use your head, in other words, or you'll literally lose your head.

Although Simonides seemed to sense this truth, some of his ancient Greek acquaintances reportedly scoffed at his memory palace mnemonic technique. When the method was described to the Athenian politician Themistocles, he reputedly quipped: "I would rather a technique for forgetting, for I remember what I would rather not remember and cannot forget what I would rather forget." More than 2,000 years later, scientists are still working on that one.

part three

SPIRIT

SELF-IMAGE

ART

ALTRUISM

LOVE

magpie in the mirror

REFLECTIONS ON AVIAN SELF-AWARENESS

When a group of German researchers announced in 2008 they had discovered that captive Eurasian magpies can recognize themselves in a mirror, many scientists were surprised. Until then, only humans, the great apes, orcas, dolphins, and elephants—large mammals with big brains—had been shown to recognize their own images, despite a slew of experiments with mirrors and other animals, including many other types of birds. Yet this particular study showed unambiguous results. Three out of five individual magpies—Gerti, Goldie, Schatzi, Harvey, and Lilly—clearly recognized the bird in the mirror as themselves.

The mirror test, as it's generally known, is straightforward: Stick an animal in front of a mirror and watch what happens. If it recognizes its reflection, it passes; if not, it fails. The tricky part is interpreting the results. There are two main problems: determining whether the animal really recognizes itself, and then figuring out what it means.

The German researchers knew this, and documented their study well enough to allay suspicion that the birds had really passed the test. They performed three different experiments with their five captive magpies. First, they introduced the birds to an open room with a dull gray plate leaning against one wall, and then replaced the plate with a mirror to see whether the magpies behaved differently in front of their reflection. Second, to gauge interest in the mirror, the researchers herded the magpies into an aviary with two connected compartments, one with the mirror and one with the nonreflective plate, and recorded how much time the birds preferred to spend in either side. Finally, the researchers marked the five magpies with a spot of colorful dye on the chin that would be invisible to the birds except by using the mirror. If the birds scratched at the spot on their own chins while gazing into the mirror, they would show a basic understanding that their reflection wasn't some other bird with something on its face. This type of "mark test" is the traditional cornerstone of mirror experiments with animals.

The first experiment clearly showed that the mirror affected the magpies' behavior. At first, all five of the birds became visibly confused. They postured to their reflection as if it were another bird and searched behind the mirror for the perceived companion. The magpie named Harvey picked up several small objects in his beak and presented them to his reflection while flipping his wings as if in courtship, and continued to display aggressively in additional trials, as did Lilly. Gerti, Goldie, and Schatzi seemed to realize the deception more quickly and quit any kind of social behavior after one or two experiences in the mirrored room.

In the second experiment, Gerti, Goldie, and Schatzi explored the mirror at greater length, often slowly moving in front of it while intently watching their reflections, and spent most of their time in the mirrored compartment, showing intense interest, while Harvey and Lilly preferred to sit in the non-mirrored side. A rift was developing between the reactions of the first three magpies and the latter two.

The final mark test was most compelling. When a colorful mark was placed on their chins, Gerti, Goldie, and Schatzi each tried to claw at themselves when they saw their reflection. Gerti and Goldie continued to scratch at their chins in trial after trial, stopping only after the mark or mirror had been removed. The only explanation for their behavior was self-recognition; when the birds were tested with a black mark that blended into their dark plumage, they appeared not to notice it, showing that they were indeed using the mirror. Gerti, Goldie, and Schatzi thus became the first three birds ever to pass the mirror test.

Because no bird had ever been shown to recognize itself in a mirror before, this experiment exceeded all expectations, with three out of five passing the test. The happy German researchers pointed out that chimpanzees, which have demonstrated the clearest evidence of visual self-awareness of any animal except humans, manage to pass the mirror test only about 75 percent of the time, even in the most productive studies. The magpie experiment wasn't meant to cast sweeping judgments of avian intelligence—even if all five birds had passed, the sample size was too small to make generalizations—but rather to show a potential ability that hadn't previously been recognized in birds at all. To that end, the study was wildly successful.

But it couldn't answer the mirror test's overall question: What does it mean, anyway? Ornithology had suddenly crossed into psychology and even philosophy. Ever since Descartes published his famous "Cogito ergo sum" revelation—"I think, therefore I am"—in the early 1600s, self-awareness has been a basic tenet of philosophy, and what some believe is an essential element of what it means to be human. Self-recognition may be the first step, but from there, the concept of self-awareness spirals into discussions of consciousness—an intuitive but slippery word that defies scientific definition despite centuries of debate. Although nobody claims that magpies are philosophers (or humans), the mirror experiment raised some new questions about avian intelligence and how it differs from our own. The implications of visual self-recognition in magpies aren't altogether clear, but the ability to knowingly admire your image in a looking glass must mark a kind of intelligence. Most animals can't do it, and that makes magpies interesting.

At least, we *think* most animals can't do it. Some scientists regard the mirror test as a flawed experiment, maintaining that it's impossible to prove a negative result because we can never really know what animals are thinking. Short of asking directly, how could we ever tell for sure? Maybe animals just don't care about mirrors enough to bother cooperating with our experiments. Because mirrors don't offer many advantages in the wild, animals might think they're boring and ignore them even if they know how they work.

And animals that have never seen a mirror might take time to get used to it, just as humans do. Children need repeated exposure to mirrors before they can comprehend the trickery of their own reflection—babies generally don't figure it out until about age two—and even adults who have been blind their whole lives can be fooled by mirrors once their sight has been suddenly restored. Although we generally take our own reflections for granted, they require some practice, even for us.

Despite these criticisms, there does seem to be a worldly divide between those who can and can't recognize their own reflection, and the German researchers showed that magpies belong in the first category, with us and just a couple of other higher animals. Interesting, indeed. But what does it mean? And why magpies?

WITH THEIR FLASHY black-and-white attire and bold habits, magpies have been familiar to humans for a very long time. They were called simply "pies," or "pyes," throughout much of English-language history. The prefix was probably added sometime in the sixteenth century, "mag" being a nickname for

"Margaret," which was used as slang for anything feminine—in this case, perhaps because people perceived the birds as idle chatterboxes.

In Europe, where magpies are most common—the thirteenth most abundant bird in Britain, according to the Royal Society for the Protection of Birds—they figure prominently in folklore and superstition. If you see a gang of magpies, according to widespread tradition, your fortune will be based on the number of individuals in the group; more precisely, two birds will invite good luck, but a lone magpie can bring down all manner of curses. Upon encountering a single magpie, prudent citizens should therefore greet it with respect by calling out "Hello, Mr. Magpie" or "Good morning, sir" to allay bad luck, or perhaps chant "I defy thee" several times in quick succession to the same effect. It also helps to salute the bird, spit on the ground, and pinch whomever you're walking with. Just in case.

The dark reputation of magpies runs deep. In Scotland, magpies foretell death and carry Satan's blood in their mouth (which has a red lining, the only drop of bright color on an otherwise black-and-white bird). In France and Sweden, the birds are regarded as thieves, probably because of their supposed habit of stealing shiny objects, especially valuable ones. Magpies were closely associated with witchcraft during the Middle Ages, along with ravens and black cats. One English folktale goes so far as to suggest that when Jesus was crucified, all of the world's birds sang to comfort him except the magpie, which was forever cursed in consequence.

By contrast, Asian cultures, particularly Chinese and Korean, have historically embraced magpies. In China, the birds are extremely popular, seen as helpful, and bring good news and joy. (This dichotomy ties in, incidentally, with the reputations of bats and dragons, which also have been reviled by

Western civilization but admired by those in the East.) Native North American tribes also tend to portray the birds positively, often as guardians or helpful messengers in traditional creation myths.

Similar-looking black-and-white magpies occupy most of Europe, Asia, and western North America. (The Australian magpie, though superficially similar, isn't closely related to its counterparts in the northern hemisphere, and there are eight species of more colorful magpies in Southeast Asia.) These are currently recognized as three distinct species: the Eurasian magpie of Europe and Asia, the black-billed magpie of North America, and the yellow-billed magpie, which replaces the black-billed in California's Central Valley. All three appear nearly identical and are so closely related, DNA evidence shows, that they can be regarded as one globally distributed superspecies.

Along with other members of the corvid family—crows, ravens, jackdaws, rooks, jays, nutcrackers, treepies, and choughs—magpies have long been thought by scientists to be among the world's smartest birds, with parrots a close second, and among the most intelligent of all animals. Most corvids are highly social, have large brains, and develop slowly—all characteristics that probably contribute to greater cognitive abilities—and magpies are no exception. When the birds' intelligence is combined with their bold, curious, and often mischievous personalities, they can impress us in surprising ways.

Consider the case of Won Young Lee, a Ph.D. student at Seoul National University in Korea who, in 2009, was helping with a long-term study of magpie breeding success on the university campus. As part of his fieldwork, Lee climbed trees to check on various magpie nests, handling the eggs and chicks to obtain measurements. Partway through the season, he noticed something eerie . . . the magpies were following him around.

He couldn't walk outdoors without a squawking, scolding bird hounding his heels. The birds didn't bother any of the other 20,000 students on campus, and only the magpies with nests monitored by Lee seemed to harass him. Even when he switched hats with a friend, the birds weren't fooled. So Lee devised an experiment. He dressed two researchers in the same clothes, instructed one to climb trees with magpie nests while the other took data on the ground, and later returned to see how the magpies would react. The birds predictably mobbed the guy who had messed with their nests and left the other researcher alone, showing an uncanny ability to recognize human faces.

Magpies are well known for taunting larger animals, especially pets. They are probably just trying to drive off a perceived predator, but sometimes they seem to consciously trick other creatures with mean-spirited mind games. One BBC documentary featured a pet magpie that loved to torment two domestic dogs by imitating the alarm call of ducks on the pond outside his house; this would invariably send the poor canines scrambling outside to chase a nonexistent fox—because the ducks often called warnings to one another when the fox passed by. Another pair of magpies once repeatedly taunted a cat along a busy country road in Britain by perching in a tree, waiting for a break in traffic, and then flying down to the pavement to lure the kitty into the road; when a car approached, the birds would flutter up at the last second while the cat scrambled to avoid becoming roadkill.

Theft seems to be a persistent personality trait. Rossini was inspired to write an opera in the early 1800s called *La Gazza Ladra—The Thieving Magpie*—and people who have an unusual preoccupation with shiny objects are said to have "magpie syndrome." This thieving reputation may be part folktale, but the birds do occasionally swipe things, often for no obvious purpose. When a magpie was caught stealing a customer's car keys at a garage in Littleborough, England, it made the *Manchester Evening News*, and also in Britain, *The Telegraph* reported in 2008 that a magpie had snatched a woman's $5,000 platinum engagement ring from her windowsill while she was in the shower—luckily, her husband-to-be found it tucked safely in the bird's nest in a nearby oak tree, albeit three years later!

One of the most intriguing behaviors of wild magpies involves their apparent habit of holding impromptu funerals. Sometimes, when a magpie finds a dead comrade, it will begin squawking at full volume, calling in all other magpies in the area, which join in an intense racket as they gather around the body. At some point, they all go quiet; there follows a period of contemplation, during which time different individuals will sometimes gently probe or preen the carcass, before each bird silently takes its leave, one by one.

Such funerals have been well documented, both with Eurasian magpies in Europe and black-billed magpies in North America. The scenario sometimes involves road-killed birds (magpies occasionally get run over while scavenging meat of other flattened animals, which can lead to pileups of carcasses on busy highways). In 2009, a researcher from the University of Colorado published detailed observations of four magpies at a funeral alongside the corpse of a fifth bird, and concluded that the birds were displaying humanlike emotions. According to his report of the occurrence, two of the individual magpies flew off and returned with grass, which they tenderly laid beside the dead bird, "stood vigil" for a few seconds, and then silently departed. After his paper was published, the researcher received a barrage of e-mails from others who had witnessed similar events. Why not? Magpies are, ironically, perpetually dressed in formal black and white. If dolphins can form friendships, rats can display empathy, and elephants mourn their dead, even bullying magpies might be forgiven for showing grief.

These days, traditionally human characteristics seem to be falling like dominoes. Magpies can recognize themselves in a mirror and apparently have a sense of self just as we do. Is it going too far to suggest that they show behavior analogous to human emotions, too?

MAGPIES DISPLAY THEIR INTELLIGENCE in many ways—thieving, holding a grudge, taunting, and even grieving—but self-recognition sets them apart from other birds. Though not every species has been tested with mirrors in a lab, many birds do encounter mirrors in today's "wild" suburban landscape, and observations indicate that most aren't able to recognize their own image.

In March 2012, for instance, a birder visiting the lighthouse at Florida's St. Marks National Wildlife Refuge noticed a male northern cardinal attacking its own reflection in the side mirror of a car in the parking lot. This wasn't too unusual; many songbirds—female brown-headed cowbirds are routine offenders—have been recorded engaging in this behavior, and the same birder had also noted a cardinal battling its reflection in almost every window of his house over the previous weeks. It was spring, and the local cardinals were pumped up on territorial hormones. But the one at St. Marks seemed incredibly aggressive. It rotated between three or four different parked cars, each time perching precariously on the rubber trim at the base of one of the front windows, where it could look straight into a side mirror. Then it would take a deep breath and launch itself at its reflection, fluttering against the mirror in a futile attempt to drive off the perceived intruder.

Others had noticed this bird already. One birder had photographed the same cardinal attacking a car mirror three months earlier, and remarked that it had kept going for at least an hour on that occasion. Another person had watched the cardinal battling against reflective bumpers as well as mirrors. This bird would even attack the mirrors of cars that had just pulled into the parking lot, flying toward them like a miniature red kamikaze as soon as the people got out.

Northern cardinals, aggressive and strong for their size, aren't known for being the cleverest of animals, but you'd think eventually the bird would learn its lesson. After beating against itself so many times, shouldn't the bird realize the difference between a real, live opponent and its own reflection? Didn't it at least wonder why it always seemed to hit an invisible barrier? But the poor cardinal continued to tilt at parked cars like a feathered Don Quixote, driving itself to exhaustion, day after day, month after month.

This happens all the time in suburban areas around the world, most often during the breeding season. Territorial birds glimpse their reflections in windows, car mirrors, and other surfaces, and reflexively try to drive away the "other bird," sometimes returning regularly for months or even years without ever realizing their mistake. This business can be particularly annoying for people who must endure the constant sound of a bird tapping on their window—and a bit creepy for some who believe that birds are souls of the dead, perhaps foretelling the imminent demise of someone inside the house.

There's not much you can do if a bird starts attacking your windows other than block the outside of the glass with something nonreflective. If, as one person recently reported, a robin starts assaulting fifteen different windows on your house, and you don't want to turn your home into a windowless cave, you just have to wait out the siege. Cardinals attack windows in North America, robins do it in Europe, and magpie-larks (no relation to magpies) are routine offenders in Australia, to the dismay of thousands of peaceable citizens.

This odd behavior suggests that cardinals, robins, and many other birds fail the mirror test. Even with repeated exposure, they never get the trick. One wonders why birds don't also attack their own reflections in calm bodies of water; perhaps they have learned about the way light behaves on a puddle, or maybe the angle of looking down at a flat surface doesn't set off the hardwired response to a rival. In any case, most birds don't seem to understand mirrors at all; they just don't realize the essential difference between themselves and other individual birds.

Which means magpies might possess an abstract capability that most other birds lack. If only we could understand what it is, we might learn something about ourselves, too.

PSYCHOLOGIST GORDON GALLUP, JR., has spent much of his life pondering the concept of self-awareness in animals. In 1970, he pioneered the use of mirrors with animals by placing a full-length mirror in a chimpanzee's cage for ten days. He watched as the chimp first reacted with alarm, then apparently began to understand its own reflection as it groomed and made facial gestures in front of it. Gallup was fascinated. Could the chimpanzee really recognize itself? To be certain, he colored different spots on the chimp's head with an odorless dye to see whether the chimp's fingers would gravitate to those spots as it looked in the mirror. The chimpanzee cooperated, and the results were published in *Science*. Researchers have been using the mark test ever since on all kinds of creatures, from those German magpies to one-year-old humans.

After his first experiment with chimpanzees, Gallup expanded. He found that all kinds of monkeys, including tamarins, marmosets, capuchins, baboons, and macaques, never passed the test, no matter how long they were exposed to a mirror (some experiments lasted years). Instead of using their reflection to groom themselves or check out parts of their genitals that they'd never seen before, as the chimps did, monkeys always acted as though they were interacting with another individual, like the cardinal at the St. Marks lighthouse.

Some chimpanzees seemed to spontaneously recognize themselves after the initial adjustment. Additional research showed that all of the great apes—chimps, orangutans, gorillas, bonobos, and humans—can pass the mirror test, although documentation has been weaker for gorillas than the other apes. This led Gallup to propose that self-awareness is present in the great apes but not in monkeys, perhaps one indication of the mental differences between the two major groups of primates.

Other researchers followed up with evidence that bottlenose dolphins, orcas, and Asian elephants could also pass the mirror test, complicating Gallup's findings. Eurasian magpies were an especially odd addition to the list, because birds have different brain structures from those of mammals. Mammals and birds diverged about 300 million years ago; we ended up with the frontal cortex, while birds have a cluster of different structures in the same region. The German researchers who conducted the mirror experiment with magpies suggested that self-recognition could have evolved independently in birds and mammals, and that the frontal cortex is not a prerequisite for intelligence. They believed that social behavior is a better predictor of mental ability than is brain structure.

Each study seemed to raise more questions than answers, partly because of the difficulty in assessing the mental states of animals without being able to communicate directly with them. So Gallup decided to shift his focus from chimpanzees and monkeys to a more familiar species: humans.

People have an interesting relationship with mirrors. Studies show that babies can't recognize their own reflections until they are about eighteen months old. Most children develop the ability by the age of two, with a few notable exceptions. For instance, some mentally disabled people never learn to recognize their own image. Self-recognition is often delayed in people with autism, and up to 30 percent of autistics never learn it at all. Patients with schizophrenia are likewise apt to respond to their reflections as if to another person. Some Alzheimer's patients also lose the ability to recognize themselves late in life.

There have been several cases of brain-damaged people who suddenly lose self-recognition, including one man who could identify other people in mirrors but not himself. The damaged area is usually located somewhere in the right prefrontal cortex of the brain—just above and behind your right eyeball—which suggests that self-recognition can be attributed to that specific region. In one captivating study, epileptic patients were each shown a hybrid photo of their own face combined with a celebrity's face while either the right or left side of their brain was anesthetized. Those with the left side of their brain "turned off" recognized themselves in the photo, but those with the right side anesthetized did not.

At about the same time that babies begin to recognize their own reflection, they start becoming aware of the thoughts and feelings of others—for instance, by showing embarrassment or trying to help a mother in distress. Gallup believed these two conditions are linked. Only by having a sense of self, he reasoned, can you make inferences about others' thoughts and actions. Thus, only creatures with self-awareness should display gratitude, deception, empathy, sympathy, humor, and associated mental states.

That's a pretty significant connection if most of the world's animals really don't have a sense of self. Perhaps the world is divided into creatures that can comprehend their own being—and infer about the experiences of others—and those that merely see others as mates or competition. (Farther down the slippery scale of consciousness, trees, amoebas, and other less-aware living things could constitute a third group.) If that's true, according to Gallup, your pet dog or cat falls into the second category, along with most birds, but magpies belong in the first—with us.

Whether the mirror test accurately measures selfhood is debatable, partly because the test allows the possibility of false negatives. One researcher showed that young children who had been marked with face rouge passed the test more often if they first watched someone else remove a spot of rouge from their own face, indicating that the mark was undesirable. Animals might likewise identify a mark on themselves without bothering to remove it. Also, many animals rely mainly on senses besides eyesight; dogs, for instance, are oriented to smell rather than sight, so might not react to a mirror even if they did possess a sense of self-awareness.

The prefrontal cortex is known to be linked to personality, prediction, and episodic memory—the brain's ability to "time travel" to recall past events at particular places and times—and probably helps make decisions about socially responsible behavior. This was demonstrated in a famous case from 1848, when Phineas Gage, an unfortunate railroad construction worker, had an iron rod driven through his skull from the back of his left cheek to the top of his head, passing behind his eyeball and through the brain en route. Though he survived the accident and recovered, his friends noticed a frightening change in his personality afterward. Overnight, Gage transformed from a well-adjusted human being into an irritable and quick-tempered man, and stayed that way. He also became inefficient and impatient at work even though he could perform tasks with the same dexterity as before. More recent studies have shown that the prefrontal cortex controls our ability to delay instant gratification in favor of a greater long-term result—something that most birds may not be able to do, at least consciously.

Magpies, with their mischievous personality, ability to recognize individual predators, and unique social behaviors that hint at emotion—such as holding funerals—are good candidates to develop the sense of self that we associate with intelligence. They have relatively large brains, comparable to apes and slightly below humans, even if the specific structures are organized differently. So why are we surprised that magpies can recognize themselves in a mirror? They may not be building spaceships, but we probably don't give them enough credit for their street smarts.

By the same logic, we shouldn't expect all birds to pass the mirror test. In 1981, a group of researchers decided to show that pigeons could do it. They expended inordinate efforts in training the birds to peck at marks on their body that were visible only in a mirror. Nobody, including the experimenters, ever argued that the pigeons could spontaneously recognize their own reflections. Rather, the study was a challenge to the mirror test, suggesting that its results should not necessarily be interpreted as self-awareness, because even animals lacking self-recognition could be taught to pass. In the end, the pigeons did appear to pass the test, but their performance showed no hint of spontaneous recognition, only rigorous training.

Gallup was not impressed by the pigeon experiment. "Training an animal to respond to marks on its body, without collateral evidence of self-recognition," he later wrote, "indicates more about the achievements of the researchers who designed the training procedures than any underlying ability of the animal."

Nothing, in other words, can replace that magic spark when you gaze into a mirror and recognize the curious face staring back, eyeball to eyeball, as your own. From Snow White's vain stepmother to Michael Jackson's haunting lyrics "I'm starting with the man in the mirror, I'm asking him to change his ways . . . ," we humans have long been fascinated by mirror images. On reflection, we might recognize some part of ourselves in Mr. Magpie.

arts and craftiness

THE AESTHETICS OF BOWERBIRD SEDUCTION

On a blistering afternoon in the Australian outback, when I stumbled into my first bower, I thought it was some kind of religious altar or maybe a practical joke. I'd been tramping through dense bush all morning and unexpectedly emerged into a fifteen-foot-wide clearing, in the center of which stood a wickerlike construction, about two feet high, resembling a small hut. It was formed of twigs woven vertically into two thick, parallel walls that created a tunnel in between, and just outside each entrance lay a pile of white stones, bleached bones, and green leaves, clearly arranged by design. The whole tidy array was surrounded by an expanse of ground so bare that I wondered whether it had been vacuumed.

As I puzzled over the curious offering, a football-sized brown bird materialized on a branch at the edge of the clearing, announcing its presence with an explosion of chattering and snapping sounds. Suddenly, it all made sense. I had wandered into the bachelor pad of Australia's winged Casanova, the serial womanizer of the avian kingdom—the great bowerbird.

I stepped back a few paces and tried to recall everything I knew about bowerbirds while this one, apparently a trusting sort, hopped down to ground level and with barely a glance in my direction set to work on his bower. He first checked over the piles of loose objects, which must have been selected and gathered with care from the surrounding bush. Head cocked, he stepped around the stones and leaves to admire his creation from various angles, occasionally darting in to nudge one of the objects with his beak like a painter correcting some tiny mistake. Satisfied, the bird then spent a few minutes weaving fresh twigs into the bower, painstakingly poking them in one at a time to strengthen the hut. I stood in quiet amazement: This was like watching a nature documentary on TV.

Bowerbirds have long been renowned for their bizarre and compelling courtship rituals. Instead of using just songs or bright feathers to attract a mate, male bowerbirds create elaborate structures to show off their unique talents of architecture and design, dedicating great energy to the task. Building the perfect bower can take ten months of each year, but it's a worthy project because females pick their mates solely by inspecting these bachelor pads, open-house-style. After mating on the spot, the female flies off to build a separate nest, lay eggs, and raise chicks by herself. Male bowerbirds can't afford to spend that much time away from their life's work, or they'll lose their seductive edge. A successful male might mate with dozens of females over the course of one season.

About twenty species of bowerbirds occupy Australia and New Guinea, and they all, uniquely among birds, display variations on this behavior. Each species has its own taste. The satin bowerbird of eastern Australia, metallic black with blue eyes, decorates its bowers with bright blue objects—berries, leaves, bottle caps, straws, ballpoint pens, plastic spoons, clothespins, anything in the right hue. The male regent bowerbird, a striking black-and-yellow bird, also from eastern Australia, paints the inside of its bower with a sticky, pea-green concoction of crushed plants mixed with saliva; females like to taste the paint when they step inside. The Vogelkop bowerbird of western New Guinea, an unassuming olive-colored bird about the size of a thrush, builds triangular huts in the rainforest and adds a mat of moss next to the entrance like a front lawn, which can cover several square meters, on which it displays eye-catching arrangements of hundreds of bright berries, beetle wings, and flowers.

The great bowerbird, from northern Australia, is the largest of the family and builds the biggest bower structures. It vaguely resembles a big, chunky robin, fawn brown all over, with scaly patterning on the back and staring, black eyes. Males have a small, muted patch of pink feathers on the nape; otherwise, males and females look pretty much alike in their drab plumage. Great bowerbirds tend to prefer green and white objects for decoration, as in the arrangement I discovered of green leaves, white stones, and bleached bones. Rocks, bones, sun-bleached dung, shells, berries, leaves, and other natural debris are typical, but human trash is also fair game. Broken glass, plastic, marbles, and nails seem to be most alluring.

One photographer documented a single great bowerbird in Queensland that had collected bits of rope, lots of broken green glass, bottle caps, lids, a plastic elephant, and a toy soldier. When a researcher tried adding colorful bits of wire to several bowers, to see if the birds would incorporate them into their displays, he stirred up trouble—neighboring male bowerbirds kept stealing the wire from one another. Though the form remains relatively constant within each species, individual birds take advantage of local resources, which can lead to fads in particular areas and in particular years. If a bunch of green plastic becomes available, suddenly green plastic will be all the rage.

Bowerbird fashion has probably been driven by sexual selection in the same way as, say, a peacock's long tail has. In the case of the peacock, females preferentially mated with long-tailed males over time, so those with short tails were eventually weeded out of the population. The same process can work with behavior. At some point in the past, female bowerbirds began to prefer males that collected precious objects, so those were the birds that passed on their genes; the more often females picked these males, the showier their displays became, in a loop of positive feedback. Bowers might be seen as part of a bowerbird's extended phenotype—a term coined by the evolutionary biologist Richard Dawkins to include not just the physical body but also any external characteristic of an animal that affects the passing of its genes. Bowers are subject to evolution just like a spiderweb or a termite mound.

Humans, too, are subject to sexual selection. Males with more possessions and creativity are generally more likely to attract women, as is true for bowerbirds. Art is one form of wealth, and, in a sense, artists send out signals that might arise from primal courtship urges. But this is a limited and clinical view of art: Surely we can appreciate artistic endeavors without the need to link them to seduction. There are many reasons to create art, including no reason at all, and some argue that true artists are those who express creativity for its own sake. It's hard to look at a great bowerbird's decorated bower, on which the bird has lavished so much time in tasteful arrangement, without seeing artistry. As I stood sweating in the Australian bush, watching this strange bird move his precious stones and leaves by millimeters in pursuit of the perfect visual pattern, I marveled at his designs. I wondered: Is this bird just carrying out his instinctive duty, or is he an artist? Is there even a difference?

THE DEFINITION OF *ARTIST* naturally depends on the definition of *art*, and art is one of those slippery concepts that seems intuitive but defies strict boundaries. Philosophically speaking, any single definition of art may be pointless or even harmful, as any box, no matter how big, could limit creativity. But there are a few things most definitions agree on.

Art must involve some kind of skill, in execution or idea, and the application of that skill to a result that may be experienced by both artist and viewer. There is nothing inherent to the concept of art that limits it to humans—although some dictionaries include this in their definitions, as art is almost always referred to in the context of people. Art usually involves some type of aesthetic sense of beauty (though beauty is also difficult to define), creativity, and imagination. It affects the senses or emotions of its audience, either toward a particular message or an open-ended reaction. Art is a form of communication.

In 2012, evolutionary biologist John Endler published a paper, "Bowerbirds, Art, and Aesthetics," in which he tackled the question head-on after completing a series of field experiments with great bowerbirds. To the extent that defining an artist is scientifically possible, Endler pursued a line of reasonable logic: Come up with a definition of art that fits most accepted standards, then see whether bowerbirds qualify.

He settled on a biological definition of visual art as "the creation of an external visual pattern by one individual in order to influence the behavior of others." In this sense, art is a signal that functions exactly the same way as any signal produced by the body, which could be viewed by an audience and could even result in mating with the artist. By this definition, both humans and bowerbirds produce art.

Endler went a step further and tried to define aesthetics in the context of bowerbird art, which was tricky because an aesthetic sense is usually attributed to appreciation of beauty, and beauty is a human standard. He cut beauty out of his definition entirely and turned to Darwinian logic. Endler argued that aesthetics involve an "exercise of judgment" among art objects, which leads to a "change in fitness"—the ability to survive and reproduce—in the artist and judge. So every time someone picks one work of art over another, life is impacted on an evolutionary scale for both artist and audience. It's an intriguing idea.

This whole argument assumes that bowerbirds can visually rank one bower over the next, which isn't by any means a given. But one of Endler's own field experiments offers evidence that they do, with a fascinating conclusion.

He wanted to understand how male great bowerbirds design their bowers to best woo a female into mating. The courtship ritual is straightforward. When a female visits a bower, she typically walks around the outside to inspect the handiwork before stepping into the avenue created between the parallel walls of upright twigs. At this point, the male, quite excited, runs to the "court" of collected objects in front of the female and frenetically picks up his most prized possessions, one by one—perhaps starting with a bright clothespin—and waves them in front of her while fanning the pink party crest on the back of his head. Instead of standing on top of the pile, which would block her view of all the objects, he angles his body off to one side, and the female can see only his head poking through the entrance. She makes her decision based on this performance.

The ritual is essentially a theater production, with the audience (the female) in its designated seat (the bower) and the performer (the male) on stage (the court). The female has no choice but to watch from a place that has been predetermined by the male. This means that the odds are shifted a bit: She may make the final decision, but he can use performance tricks to influence her judgment.

Endler found that the layout of objects around the bower is

far from random. Male great bowerbirds place larger objects progressively farther away from the entrance, creating a forced perspective—from the female's viewpoint, looking out, they all seem the same size. This is definitely on purpose. When the objects are experimentally rearranged, male bowerbirds reposition them within a few days, and, while working on the array, they will often walk into the bower and peek out as if imagining what the female will see. The perspective creates an illusion that objects farther from the bower are smaller than they really are, and, correspondingly, that the male beside them is larger than he really is.

Forced perspective has been used by human artists for centuries. Ever taken the classic tourist photo where you stand with a mountain in the background and reach your hand over it, seemingly resting your fingertip on its summit from the camera's viewpoint? That's an extreme example, in which the mountain looks to be two inches tall. Michelangelo used forced perspective on his statue of David in a more subtle way—as the sculpture was meant to be viewed from below, he enlarged the torso and head slightly so that they wouldn't seem to be dwarfed by the feet (the effect is obvious when viewed from the side). The Greeks made their columns narrower at the top so they would look taller. Architects of the Cinderella castle at Disneyland designed tiny turrets for the roof that would fade into the distance, making the castle appear larger than life. The illusion depends on an audience observing from a predictable viewpoint, like a female bowerbird hunkered inside the bower. Male bowerbirds have learned the trick.

When Endler analyzed the mating success of each individual bowerbird in his study, he discovered that the males who got the most action had created bowers with the most regular geometry, best forced perspective, and deepest visual illusions.

This showed that females could distinguish between visual patterns of varying quality. And because these choices affected mating success for both males and females, the study showed that bowerbirds, according to Endler's definition, do have an aesthetic sense.

Other animals build structures, some even decorated. Many birds camouflage their nests by using material matching their surroundings. The great crested flycatcher of North America habitually hangs snakeskins from its nest, which is thought to ward off predators. Some spiders add special silk decorations to their webs that may attract insects or deceive birds. By Endler's definition of art—creating an external visual pattern to influence the behavior of others—all of these animals qualify as artists. But in each case, these structures also have a separate purpose: Nests are shelters; webs catch food.

There's nothing wrong with functional artwork; just about anything with a practical use can be made artistically. Some would say that good design of everyday objects—chairs, computer graphics, clothing—is an artistic discipline. But what we generally think of as art, what we call "fine art," most likely to be found in a gallery or museum, usually doesn't have a function besides visual communication. The primary goal is to influence the viewer's emotions and behavior.

By this narrower, art-for-art's-sake definition, Endler believes that bird nests, spiderwebs, and all other animal structures are ruled out—with the singular exception of bowerbird bowers, which have no physical function. Only bowerbirds and humans, he says, create and exhibit objects with the sole purpose of modifying the behavior of their viewers. And this, I believe, leads to a tantalizing conclusion, that if a well-designed bower is more like a Picasso painting than a piece of furniture, then the bowerbird is no mere carpenter—he is Picasso himself.

HUMANS BEGAN CREATING ART at least 40,000 years ago, decorating European caves with realistic paintings that are still visible today. We don't know how or why our early ancestors felt the need to make representations of their surroundings, but the stenciled images of animals and human hands—interspersed among a lot of abstract doodles—are haunting. Cave paintings remind us that art has been part of the human condition for a long, long time.

Just how we became so fascinated with art is the subject of major dispute among historians and biologists. Its origins are hidden in the early mists of painting, writing, music, language, dance, theater, and religion, which overlap increasingly as we plumb the depths of history. These disciplines collectively express the human brain's capacity for abstract thought, representing ideas with shapes, sounds, and movements—an advanced mental ability that first allowed us to develop modern culture. All art is abstract in this sense, and it requires processing power from any artist.

Some argue that art developed as a by-product of our powerful brains, others say that art gave us an evolutionary advantage and was born of strict Darwinian selection, and still others see art as a product of social culture. However you look at it, visual art has been found in virtually every human society for which records exist, qualifying art as a human universal. But is it just for humans?

An art professor from New Zealand, Denis Dutton, makes the case in his recent book *The Art Instinct* that art evolved in humans the same way that peacocks got long tails, through pure natural selection. Art, he says, helps people find a mate. Quality paintings and sculptures convey status and, if they are presented by the artist, display personal skill. The same is true of other arts: Music is the language of seduction, dance is a courtship ritual, and so on. We value art because it represents wealth and achievement. Art is useful, gives an advantage to those who practice it, and so, logically, has become integrated into the very core of human being.

By making this argument, Dutton allows the possibility that art isn't restricted to humans. Because the force of natural selection acts on all life-forms equally, there is little reason to suspect that artistry would evolve only in people.

The title of his book reflects this shift in thinking. Art is usually considered a creative process, but Dutton suggests that we have less control over it than we realize, that we are driven to make art by innate tendencies that have evolved over aeons. As it happens, this is probably the most compelling argument *against* artistry existing in bowerbirds because it implies that the birds aren't creative, but that they make their designs entirely by instinct. All great bowerbirds build walls of twigs decorated with the same assortments of loose objects. Because the form doesn't vary much from one bird to the next, it can hardly be considered imaginative.

But even bowerbirds show glimmers of inventiveness, as the geography professor Jared Diamond demonstrated in the early 1980s, years before he wrote *Guns, Germs, and Steel* and other books that made him famous. While doing fieldwork in remote New Guinea, Diamond discovered a previously unknown population of Vogelkop bowerbirds, the species that builds elaborate triangular huts surrounded by mats of moss and colorful objects. This new, isolated population looked and behaved the same as previously studied birds, but seemed to prefer brown and black decorations instead of bright colors—a sort of goth subculture. Diamond wondered how they could have learned new tastes, and decided to conduct an experiment to find out.

Out in the rainforest, Diamond scattered seven different colors of poker chips near bowers to see whether the birds would incorporate the novel objects into their displays. This was nothing new; bowerbirds in other areas readily snapped up human trash. Diamond really wanted to test whether individual birds have personal visual preferences, and he supplied the poker chips just to ensure that all the bowerbirds had access to the same materials.

He found that most traditional Vogelkop bowerbirds went crazy for the poker chips, especially the bright blue and red ones. They were so popular that neighboring males frequently stole them from other birds' displays, forcing Diamond to number them just to keep track of which chips went where. But the birds of the new population, which liked dark berries and rocks, never touched the poker chips. It seemed that they had developed a new aesthetic standard, while in all other ways remaining alike to the other birds. Because the populations were now separated by their visual tastes, with females choosing males having either bright or dark bower decorations, the birds had become reproductively isolated from each other. The implication was startling: If this trend continued, aesthetic preference would eventually lead to two different species of bowerbirds. Instead of evolution driving art, art was driving evolution.

Also, individual birds within each population differed significantly in their color preferences, and adult males built more complex bowers than younger males. Diamond observed that because males often stole objects from one another, they had lots of chances to size up their neighbors' technique and might pick up some tips. Young male great bowerbirds sometimes work collaboratively on a single structure before striking out on their own, and females may travel in pairs or small groups when sizing up the local talent. In Diamond's view, the aesthetic taste of bowerbirds, though based on instinct, is partly learned. Which means that bowerbird aesthetics are culturally transmitted, much like human art styles.

So we have people like Denis Dutton arguing that human art is *more* instinctive and others like Jared Diamond saying that bowerbird art is *less* instinctive. Even in art, the perceived gulf between people and animals is being eroded at both ends.

Not that the gap doesn't exist. Dutton's theory that art is strictly a survival benefit goes only so far. It may help explain why rich people hang million-dollar canvases in their mansions and why we listen to music in dance clubs, but it's more difficult to relate Darwinian selection to, say, the illustrations in textbooks or the tunes in your personal playlists. As a type of communication, art may convey many messages besides seduction and status. And bowerbirds are hardly painting Renaissance masterpieces; only humans create representational art of any kind. But the parallels may be deeper than we thought. Given another million years to refine its technique, one can only guess what the bowerbird might come up with.

IN 1872, A CAPABLE ITALIAN BOTANIST, Odoardo Beccari, who rubbed shoulders with Charles Darwin once or twice, was exploring a New Guinea jungle when he stumbled across a small hut. "I had just shot a small marsupial as it was running up a tree," he later reported, "when turning round close to the path, I found myself in front of a piece of workmanship more lovely than the ingenuity of any animal had ever been known to construct. It was a cabin in miniature, in the midst of a miniature meadow, studded with flowers."

Beccari—probably the first European naturalist ever to lay eyes on a bower in New Guinea—took it for a construction of one of the local tribes, though he had no idea what it was for. Ceremonial offering? Children's plaything? The tiny hut was baffling. When he eventually caught the bowerbird at its "cabin," Beccari became so awed by the structure that he spent several days observing and sketching it with crayons, noting in detail the various architectural features and decorations.

This isn't the only time in history that someone has mistakenly attributed the work of animals to humans. There is the story of a classic art hoax in Sweden when, in 1964, amid the post–World War II explosion of abstract painting, a mischievous journalist persuaded a zookeeper to give oil paints to a four-year-old chimpanzee named Peter. After the ape learned to smear the colors onto canvas with a paintbrush, the journalist entered four of Peter's best works in an art show at a museum in Göteborg, Sweden's second-largest city, under the French alias Pierre Brassau. The paintings were displayed alongside those of other European artists without any indication of their true provenance, and art critics gave them nearly universal praise. One reviewer, who panned the rest of the show, wrote, "Pierre is an artist who performs with the delicacy of a ballet dancer." (A dissenting critic aptly suggested that "only an ape could have done this.") One of Peter's paintings sold for $90 (about $700 today) to a collector. When Brassau's true identity was revealed, the critics were even more impressed—and the chimp's fame spread worldwide.

Art hoaxes—as well as the plethora of artistic paintings by orangutans, elephants, gorillas, and other animals that are being sold with full disclosure by their owners—highlight the difficulty of defining art strictly by appearance. If we can't even tell whether a particular object has been made by a bird, a chimpanzee, or a human, then who are we to judge what is art and what isn't? It's just too hard, and probably hypocritical, to limit art to people. Humans do not seem to have a lock on creativity.

The best definition of art may just be the simplest: Art is whatever an artist says it is. If someone creates in the spirit of artistry, then so it shall be considered by everybody else. However circular this logic may seem, there is often no way to know what went into a piece without guidance from its creator—which is why, in art gallery exhibits, viewers often spend as much time reading the tiny informational placards as appreciating the artwork on display. We want to know who made it, how they did it, and what it means. If beauty is in the eye of its beholder, then art, by definition, is in the intention of its maker.

Unfortunately, we don't know what goes through a bowerbird's mind while he builds and decorates his bower. Does the bird derive satisfaction beyond the occasional female stopping by? Does he think of himself as talented? If we could ask, would a bowerbird call his own bower art? Because we can't ask, we're not sure how we should respond to their curious structures.

We do know that bowerbirds are quite intelligent. If any birds were to channel the abstract thought necessary to make art, bowerbirds would be good candidates. They are most closely related to the corvid family, which includes crows, ravens, and jays—perhaps the smartest birds on the planet—with boldly curious personalities to match.

Many bowerbirds are excellent vocal mimics, traditionally a hallmark of intelligence. The MacGregor's bowerbird, which inhabits mountainous parts of New Guinea, has been recorded imitating the sounds of waterfalls, pigs, and human speech.

Some species also use tools, appropriating sticks as brushes to paint the insides of their bowers with chewed-up plant goo.

They have larger brains than other similar-sized birds, and research has shown that the bowerbird species with the most complex bowers have the largest brains (and the males, who do all the building and decorating, have slightly larger brains than the females).

It does seem a stretch to suggest that bowerbirds consider themselves artists the way we do, given their overt and overwhelming motivation of seduction; the birds' attention is probably directed more at passing females than creative immortality. So the art-by-its-maker definition would rule out bowerbirds. They probably don't see art as some higher discipline. But until very recently, neither did we. Dictionary definitions of art have changed over time, and it's only within the past four hundred years or so that art has begun to stand for something other than straightforward craftsmanship. Throughout most of history, people valued art because it was useful; artists were good at making things, and artwork was defined by its inherent quality. Even during the Renaissance, celebrated painters were often viewed more as skilled craftsmen than intellectuals.

The first copyright laws weren't enacted until the 1700s, and before that, few artists had notions of creative ownership. In earlier times, copying artwork was good exercise. Those with the skill to do it were regarded with the respect of an original artist, because art and craft were the same. But that has changed. Today, forgers are thrown in jail, and a Monet may sell for tens of millions, while a forged one is comparatively worthless—even if the most knowledgeable experts can hardly tell the difference. The definitions of art and skill have separated so that we now value reputation as much as talent, concept as much as execution.

Bowerbirds are yesterday's craftsmen, spending their lives honing skill and technique to achieve the perfect bower. A few hundred years ago, that would have satisfied every definition of human art, but today it's not good enough; because the birds don't sign their names or explore themes, they fail to qualify as artists. Honestly, I'm not sure whether this distinction reflects better on us or on them.

Sometimes I feel a little sorry for bowerbirds. They have become so devoted to making a good impression that they don't have anything left for the females they want to impress. Male bowerbirds, whether struggling artists or crafty seducers, are lifetime bachelors, so absorbed in their work that they will never raise their own children. The perfect bower leaves no time for anything else.

fairy helpers

WHEN COOPERATION IS JUST A GAME

Whenever the temperature climbed above 110 degrees Fahrenheit, which happens often in northwest Australia, I would retreat into the walk-in refrigerator at Mornington Sanctuary, pull the heavy door shut, turn out the lights, and think about purple-crowned fairy-wrens. Never mind how the dainty birds survive in an environment so extreme that even Sir Sidney Kidman—Australia's famous cattleman, who, in the early 1900s, turned five shillings and a one-eyed horse into ownership of 3 percent of the entire continent—couldn't handle the area. (He cited "natives spearing the cattle" and "the precipitous state of the country" when he abandoned Mornington; no fridge back then.) I was rather more impressed with the fairy-wrens' social habits. In the loneliest corner of the scorched outback, where groceries are delivered by bush plane and solitude lies like a hot blanket, miniature soap operas play out every day. You just have to know where to look.

Fairy-wrens, with the body and bounce of a Ping-Pong ball fixed with a pencil-sized tail, make up for their diminutive stature with blatantly colorful attire. There are fourteen species of fairy-wrens overall, nine of them in Australia, all of them visually sparkling. Males of the purple-crowned variety, which is endemic to the arid northern interior, sport a candy violet hairdo, black cheeks, and a long, expressive tail that shimmers as blue as the sky. Females are more demurely patterned with slaty crowns and a reddish cheek patch broken by a white eye ring. But it's their cooperative personality that most fascinates fairy-wren admirers.

A hundred yards from my icebox retreat, one particular

female purple-crowned fairy-wren could reliably be found industriously tending her brood. As fairy-wrens go, she was relatively hardworking. Along with a studly mate, she'd already raised two healthy nestlings who had left the nest a couple months earlier. Now she was in charge of a second nest, a tightly woven ball of grass and leaves with an entrance hole on the side, protected by a stalk of spiky pandanus shrubbery along a small creek. But her earlier young, now grown to adult size, hadn't yet left the territory. Instead of dispersing, as most birds do after fledging from the nest, they hung around and helped their parents raise the next brood, bringing extra food to their younger brothers and sisters. That was where things got interesting.

I knew this because I'd spent countless hours—days, weeks, months—carefully watching them as part of a long-term study of the birds' habits. I recognized by sight every individual fairy-wren along a ten-kilometer stretch of creek, which amounted to one sprawling family of a hundred siblings, uncles, cousins, grandparents, and occasional newcomers, all packed into forty well-defined territories. It was like a suburban street, forty houses long, in a neighborhood where everyone knew one another.

On this particular street, neighbors sometimes pop in to take care of one another's kids, an unusual habit among birds. Only a very few birds, somewhere between 3 and 8 percent of the world's total species—including acorn woodpeckers, Florida scrub jays, and groove-billed anis, among others—will ever voluntarily tend another's young. This kind of cooperative nesting presents an evolutionary enigma at first glance. Why would any creature, generally assumed to place its own interests first in order to survive and reproduce, willingly help another, at cost to itself, with no obvious benefit?

The longer I watched fairy-wrens, the more I became fascinated by this question. I read up on cooperative behaviors in all kinds of animals and discovered that wild creatures don't always behave as selfishly as you might think. Vampire bats form buddy systems, regurgitating blood for each other when a pal goes a night without a meal. Dolphins push sick or injured individuals—occasionally even of other species, such as seals and humans—to the surface to breathe. Lemurs care for unrelated infants. The list goes on and on; it seems that people aren't the only ones who help one another from time to time.

Even within the realm of birds there are many examples of cooperative behavior. When a bird sees a predator, it will often give an alarm call that alerts others in the area—sometimes of different species—to the danger, even though doing so attracts attention and presumably increases the whistle-blower's chance of being eaten. Some birds that feed in flocks, such as quail, habitually post sentries to stand watch while the others eat their fill. Why take the extra risk? If every bird was completely selfish, none should be willing to protect others while placing itself in harm's way.

The literature on human altruism—good deeds performed at cost to oneself with no expectation of a returned benefit—is just as interesting. In terms of playing nice, we may not be much different from other animals, and we may have less control over our actions than we like to think. It's possible that cooperation itself can be explained by mathematical theories, with wide-ranging implications for criminals, cheating, the Cold War, the Golden Rule, forgiveness, cancer research—and, yes, fairy-wrens.

Cooperative nesting is the ultimate example of altruistic behavior, as it so clearly helps another individual affect the next generation. By tending one another's nests, fairy-wrens seem to violate a basic principle of Darwinian evolution lately championed by the author Richard Dawkins—that the need to pass on your genes supersedes all else. If reproduction is the primary goal, then it seems illogical to help raise someone else's kids while delaying your own.

Geography gives a basic hint to the puzzle. Of the few birds that do nest cooperatively, many of them are concentrated in Australia and Africa, and a lot of those birds, including fairy-wrens, live in scrubby habitats or grassy savannas. An unpredictable environment, such as a savanna that experiences sudden droughts and rainy periods—or, say, a stock market investment business—favors diversified team efforts. On a finer scale, places where prime territory is limited may foster a system in which young birds must stay with their parents until a vacancy opens up elsewhere, effectively paying rent until they can strike out on their own.

But geography can't entirely clarify the cooperative behaviors of fairy-wrens, which sometimes have nest helpers and sometimes don't. Scientists are interested in the evolution of cooperation because it seems counterintuitive; efforts to explain selfless behaviors usually involve some ultimate benefit with the idea that all behaviors, in the end, are actually selfish. The question becomes whether pure altruism even exists or whether all "nice" actions calculably benefit the do-gooder. Let's be honest: When humans are nice to one another, it's often for an ultimately selfish reason. Fairy-wrens might not be much different.

THE MOST OBVIOUS EXPLANATION for cooperative nesting is that helpers are usually relatives. In terms of furthering your own genetic legacy, a brother is the same as a son: Both share 50 percent of your genes. Assuming that they are closely related to the true parents, fairy-wren helpers are reaping the genetic benefits of raising children without bothering to have kids of their own, sort of like an older brother babysitting his younger sibling while their parents go out to the movies. It's a job, but it does end up protecting at least some of the helpers' own genes.

The Mornington study showed that this is mostly the case for cooperative purple-crowned fairy-wrens. When researchers examined the family trees of helpers, they found that 60 percent of them live with both of their parents and 90 percent with at least one parent. Helpers bringing extra grub to a nest are usually feeding their own younger brothers and sisters. The birds are generally monogamous, so eggs in any given nest can be assumed to contain the genes from the two adults—and additional helpers—who attend it. DNA testing confirms these relationships.

It seems kind of obvious that you'd want to help your relatives more than strangers, which is technically called "kin selection." Familiarity is a complicating factor; you're more apt to aid a friend than a stranger even though neither shares your genes, and family is generally most familiar because you live together. Still, you'd probably include a distant family member in your last will and testament before a complete stranger. And you might prefer to donate a kidney to a family member over a good friend. As they say, blood is thicker than water. Most examples of apparent altruism in animals take place between genetic relatives.

Accordingly, scientists have found that fairy-wrens are more likely to help a nest with chicks that share their genes, and the fewer genes they share, the less help they give. The British geneticist J. B. S. Haldane had a firm grasp on this concept decades ago; when asked if he'd sacrifice his life for a drowning friend, he quipped, "No, but I would to save two brothers or eight cousins"—referring to the fact that each brother shares 50 percent of your genes and each cousin only 12.5 percent.

But some purple-crowned fairy-wren helpers aren't related at all to those they support. Young, dispersing birds sometimes drift into an area, somehow convince a pair of completely unrelated adults to take them in, and begin helping those adults feed their chicks. Such cases are particularly interesting because they can't be explained by kin selection. If those helpers aren't gaining any genetic benefit, then why are they being so generous?

TO UNDERSTAND FAIRY-WRENS and cooperation in general, it can be helpful to forget that they are birds, and instead treat fairy-wrens as generic, logical beings in a strategic competition. Imagine that survival is just a game in which the birds are individual players. They can decide either to cooperate with one another or not in various situations, and each of those decisions will affect their ultimate success. To win the game, with the highest chance of passing on their own genes, the birds must choose a perfect strategy, so that when they cooperate with one another, they score more points than they would if they had decided to strike out on their own—and they don't expose themselves to unnecessary risks when cooperation doesn't pay off.

Looked at this way, cooperative nesting in fairy-wrens can be distilled into a problem of game theory, the study of strategic decision making. This assumes that there is such a thing as a perfect survival strategy for fairy-wrens, that some strategies are better than others, and that real life can be represented as a logic puzzle at all. But game theory, which has been well studied by contemporary mathematicians, can tell us a lot about cooperation, from global wars to cancer cells, and might illuminate bird behavior as well.

Logically, the decision to cooperate or not can be trickier than it sounds. Sometimes, short-term rewards are weighted against cooperation even when working together would pay off better. This is illustrated by a classic strategy problem known as the prisoner's dilemma.

Imagine that you have been arrested for robbing a bank along with a close friend. The police lock you and your accomplice in separate cells to await trial and then give each of you a choice: Stay quiet, or testify against your partner in hopes of a deal. You have no way of knowing what your buddy will do, but the police are perfectly straightforward. They advise that if you both stay quiet, you will each receive one-year sentences. If you each betray the other, you will both get three years. And if you rat out your partner and he protects you, he'll get ten years while you go free.

The best overall outcome is for both of you to stay quiet—in which case you'll both be sprung in a year. But you're not sure if your friend has your best interests at heart. By protecting him, hoping that he'll do the same for you, you risk the worst possible sentence for yourself. And you realize that the odds are stacked in favor of betrayal: If you stay quiet, you'll get either one or ten years, whereas if you defect, you'll get either zero or three years. Staying quiet would assume an average sentence of 5.5 years, but testifying against your partner would give you, on average, 1.5 years. Because you are selfish and logical, you betray your friend—and, for the same reasons, he betrays you. Instead of each receiving one-year sentences, you both serve three.

This is a famous problem. The situation was first described in 1950 by two mathematicians in an analysis group to the U.S.

armed forces, both experts in game theory. They realized that certain situations are mathematically stacked against cooperation, even when mutual cooperation would produce a better outcome. As long as each player in a game is selfish and logical, two opponents will not necessarily work together toward the best possible result.

They initially framed the dilemma in terms of military strategy, which would prove remarkably prescient. In 1950, the United States and the Soviet Union were just beginning an uneasy truce with each other. The United States had dropped nuclear bombs on Japan five years earlier, winning Japan's immediate surrender, and the Soviet Union had exploded its own first nuclear device during the previous year. Those two American mathematicians may have had an inkling of the four-decade Cold War arms race that would follow.

According to some political scientists, the Cold War was one big prisoner's dilemma. Each side had two options: arm or disarm. If both sides disarmed, nobody would spend money or get hurt—clearly, the best outcome. If both sides armed, each country would sink billions into a nuclear program instead of domestic projects, with the added possibility of mutual destruction. If one side armed itself while the other disarmed, though, the result would be immediate superiority. From either perspective, it was better to continue the arms race even though cooperation could have prevented the whole ridiculous, scary stalemate.

The dilemma crops up in lots of other situations, too: price and advertising wars between companies, use of performance-enhancing drugs in sports, and even the application of makeup by women. All of these situations fulfill, at least conceptually, the mathematical conditions of a prisoner's dilemma in four possible scenarios: (1) Betraying your opponent while he cooperates pays better than (2) both of you cooperating, which pays better than (3) both of you betraying each other, which pays better than (4) cooperating while your opponent betrays you.

It shows how purely logical beings can choose not to cooperate even when mutual cooperation would pay off better for everyone involved. To some degree, the prisoner's dilemma confirms the general idea that everyone is selfish to the point of self-destruction. The dilemma predicts, given a certain set of conditions, that individuals will decide to thwart each other in an attempt to get ahead. Much like the tragedy of the commons, a related social dilemma in which members of a group deplete a shared resource for selfish reasons, the prisoner's dilemma focuses on the fact that what's best for a group isn't necessarily best for any given individual.

When two individuals work together, at least one of them usually makes a short-term sacrifice. In fairy-wrens, helpers forfeit their own reproductive efforts to feed the nestlings of older adults. Such sacrifices probably lead to long-term benefits that outweigh any up-front costs of cooperation, but, assuming the birds are logical and selfish, can they really see that far into the future?

The trouble with the prisoner's dilemma is that its conditions are rarely met in day-to-day life. Any given interaction between two individuals is usually not a one-off thing; you're likely to meet your opponent again someday. If you burn him now, you might be sorry later. What goes around often does come around. This fact alone is enough to foster strategic cooperative relationships, as demonstrated by American political scientist Robert Axelrod in the 1980s.

Axelrod became interested in a game version of the prisoner's dilemma in which the situation is repeated many times in succession, with two opponents choosing in each round whether to cooperate or defect—a game that more closely mirrors real life. By the mid-1970s, more than 2,000 scholarly papers had been published about this one mathematical problem, many of them expounding on various possible strategies. Axelrod decided to host a tournament. Academics from all over the world entered their programmed algorithms, each one describing a slightly different strategy, and they all battled it out in a logical elimination until just one remained.

Many of the programs were quite complicated, but the winner turned out to be the simplest of all. Called Tit for Tat, its logic was unassailable: Cooperate at first, then, in each successive round, do whatever your opponent did in the previous round. This was interesting because in a game known to reward selfish behavior, the winning strategy played nice and punished an opponent only if the opponent didn't cooperate.

Axelrod held the tournament again the next year, and Tit for Tat won again—and again the year after that. It was eventually defeated only by multiple programs entered together that had been preprogrammed to recognize one another and sacrifice themselves to boost one overall winner, in a sense subverting the rules of the game.

When you think about it, Tit for Tat makes sense. It's the prisoner's dilemma version of "an eye for an eye," or, more optimistically, the Golden Rule ("Do to others as you would have them do to you"). If everybody cooperated all the time, the situation would be idyllic but highly unstable: anybody could saunter in at any time and take advantage. But if everyone always betrayed each other, nobody would gain anything. Stability lies somewhere in the middle.

Reviewing the best strategies in his tournaments, Axelrod found four main predictors of success: (1) The strategy must be nice; it won't cheat before its opponent does. (2) If its opponent cheats, the strategy must retaliate; otherwise, it will be walked over. (3) But it must be forgiving; instead of holding a grudge, the strategy should revert to being nice after retaliating. (4) The strategy, counterintuitively, must not be jealous; it must not score more points than its opponents in any given round. This last condition illustrates the essential difference between the single-round prisoner's dilemma and its repeated version: If you're interacting just once, the best strategy is to betray your opponent, but it's best to play nice over the long haul.

As Axelrod found out, it's easy to start generalizing these results with real human and animal behaviors, and he wrote a book about it. *The Evolution of Cooperation* explains how nice strategies, like cooperative nesting, often win in the long term, and offers evidence for how those behaviors might have evolved through natural selection. Whether or not do-gooders are acting out of the generosity of their own hearts, Axelrod argued, they're ultimately being nice to get something out of it.

This can be true even if you don't expect a good deed ever to be returned. Think of it this way: The cost of being nice in any given interaction is small, but the cost of burning someone might be huge. So, logically, even if 90 percent of your kind actions are never directly returned, those that are will more than compensate.

You can extend this line of reasoning to answer all kinds of questions about why humans and other social creatures are generally kind to one another, why we are horrified by violence, and why we cooperate. There will always be those who try to take advantage—any population without defectors would be unstable because it would invite exploitation—but, generally, it pays to be accommodating. If we weren't so nice, anarchy would ensue. But there's a less rosy side. If we cooperate only for our own selfish reasons, then does true kindness even exist? Is there such a thing as real charity? Scientifically, altruism is nearly impossible to prove, and the concept is hotly debated. Ethicists cringe at the thought that all good human behaviors may be self-motivated. But those who study animals tend to accept this view more readily, explaining away any indication of altruism as an evolutionary benefit.

Which brings us back to fairy-wrens. Their cooperative nesting habits are often cited as an example of altruism in the animal kingdom, along with other seeming acts of kindness. But the bird world doesn't really work that way. A bird that helps feed another bird's nestlings must be doing it for an ultimately selfish reason. In the end, its own survival must benefit.

THE MORNINGTON STUDY found that unrelated helpers are probably motivated by the prospect of inheriting a good territory. Because the supply of waterside habitat is limited, almost all of it is occupied by dominant adult fairy-wrens. Sometimes a young bird can carve out a place for itself only by joining an occupied space, paying rent by helping the current owners raise babies until they either die or move on. It's a win-win agreement because the adults get extra help feeding their nestlings, and the helpers get a chance to inherit a nice place to live.

It comes as no surprise that purple-crowned fairy-wren nests with helpers tend to be more productive. Yet, in a closely related species, the superb fairy-wren, which uses a similar system of nest helpers, researchers have been unable to show that the extra help translates into healthier chicks. Study after study has found no difference in fledging success between nests with and without helpers. Scientists could only scratch their heads and wonder whether their data were correct; if helpers didn't increase nesting success, why did dominant adult fairy-wrens allow young birds to squat on their territories?

Superb fairy-wrens have a different social system from that of purple-crowned fairy-wrens. Males pluck flower petals and display them to females for courtship, and females often sneak away before dawn to mate with other males. Rampant promiscuity means that, unlike purple-crowned dads, male superb fairy-wrens are often unrelated to the nestlings in their own nest, and so are the helpers (unless, as sometimes happens, the helpers mate secretly with the female). All this loose living makes you wonder what motivates superb fairy-wrens to breed cooperatively. Maybe helpers are doing it solely to inherit territory rather than to protect family. But are they even really helping?

When a team of researchers led by an Englishman, Andrew Russell, compared the fledging success of young superb fairy-wrens from nests with and without helpers in 2007, there was no measurable difference; although chicks in nests with helpers were fed an average 19 percent more food, they weren't any healthier. Earlier studies had reached the same conclusion. Russell couldn't make sense of this result, but he had a theory. If the chicks weren't benefiting from extra attention, maybe their mothers were.

He carefully measured eggs in a variety of nests and found that eggs were more than 5 percent smaller in nests with helpers than in those tended by just two adults. Smaller eggs also contained a smaller proportion of yolk and nutrients. When females could rely on a bigger family to raise their chicks, it seemed that they put less energy into their eggs. At hatching, those chicks emerged wimpy and underweight. But with additional fairy-wrens bringing them extra food, scrawny chicks grew faster, and by the time they left the nest, they'd caught up to normal fledglings with two parents.

Then, critically, Russell looked at the long-term survival of the mothers. Adult female fairy-wrens who lived with just one male had a one-in-three chance of dying within one year. But for females with helpers, those odds were reduced to one in five. Russell's results made the cover of the journal *Science*. In superb fairy-wrens, cooperative nesting doesn't really help the kids. It's the moms that soak up all the benefits.

IN THE 1980S, Martin Nowak, a graduate student in mathematics, became so fascinated by Robert Axelrod's classic prisoner's dilemma studies that he dedicated his thesis to the subject. Nowak wanted to beat the Tit for Tat strategy. What if, instead of crafting a single unchanging algorithm, he could grow a winning strategy organically through generations of breeding?

Nowak, working with his Ph.D. adviser, Karl Sigmund—who was one of the first to apply mathematical game theory to evolution—modeled a population of creatures, each with different random initial strategies of cooperation. Then he let those creatures interact in prisoner's dilemma situations over time, mirroring the rounds in Axelrod's tournaments; each one would always remember previous rounds. But Nowak added some new rules. He introduced strategic mutation, as you'd expect in any population of animals. He also forced occasional errors so that each strategy would sometimes cooperate when it meant to defect, and vice versa. Finally, he modeled selection: Successful strategies reproduced, while unsuccessful ones died out.

At first, with no history of interactions, defectors dominated. But after just a few generations, the Tit for Tat strategy suddenly outcompeted the betrayers. Tit-for-Tatters ruled for another dozen generations, but then, in an unexpected stroke, they gradually began to be replaced by a mutated version that Nowak called "Tit for Tat with forgiveness"—a strategy that copied its opponent's actions but occasionally cooperated even when its opponent defected. Forgiveness tended to score more than simple retaliation, so the population swung toward nicer strategies until virtually all individuals were cooperating all the time, in stark contrast to the beginning of the model. At that point, things became so unstable that when a few mutations introduced individuals who preferred betrayal, the cooperators were overtaken and the cycle reset.

Nowak's effects became even more striking when he added the element of reputation. Actions could now take into account not only one's own past but also the pasts of others. The population trended toward cooperation because it paid to be nice to the ones most likely to return the favor. When it reached a certain critical point, a few uncooperative strategies still took over and the cycle began again, but this time with a twist. Sometimes, clusters of truly cooperative individuals would emerge that, by interacting mostly with one another, could not be brought down with any amount of outside betrayal.

Nowak's model demonstrated that different strategies may succeed, depending on circumstances. Nobody can be nice all the time. But it also showed a fascinating rise of cooperation from the dry dust of random math, and suggested that cooperative behavior is not only beneficial but perhaps part of evolution.

Complex societies just can't function without cooperation. If nobody ever cooperated, we'd all still be single-celled organisms swimming in the soup—and, it should be noted, even unicellular slime molds can act socially. At some point, two cells had to get together, each sacrificing a bit of its own freedom to form something more ordered.

Everywhere he looked, Nowak saw examples of cooperation. He delved into the evolution of language, which he perceived as a bunch of people cooperating. He turned his attention to the mathematics of cancer, which he viewed as a bunch of cells not cooperating. The more he thought about it, the more he became convinced that cooperation was no mere feel-good behavior; it must be vital to life itself. His book *SuperCooperators* argues that cooperation should be considered a third tenet of evolution, right up there with mutation and natural selection. And he doesn't leave it there. He believes that we're all playing one large, strategic game against our future descendants, and that we'd better start cooperating with them or they won't have much of a world to live in. No model can give us the definitive answer for the end of that game.

WHEN THE BLAZING AFTERNOON melted into a calm evening at Mornington, I set out down a graded dirt track with a lawn chair, a field notebook, and a pair of binoculars. The temperature pushed a hundred degrees even near sunset, and heat wafted off the ground in invisible waves while I meandered past wallabies and fat-bellied boab trees. I marveled at Australia's red dust, which eventually clogged up every crevice of my shoes. My favorite pair of purple-crowned fairy-wrens lived about a half-mile from the station off that dirt track.

As soon as I reached their nest, I set up my chair on a grassy bank with a good view at a distance that wouldn't interfere with the birds' habits. They were used to me by now, but I respected their space. This evening, I hoped to further, by one tiny notch, our knowledge of cooperation in fairy-wrens.

Nest watches help researchers measure exactly how much effort each bird puts into its nestlings. This particular group contained four birds, all attending the same nest with several five-day-old babies inside. There was an adult male—obvious by his purple crown—along with an adult female and two younger helpers. The female and helpers were harder to tell apart, so I had to rely on unique color bands on their legs, which were sometimes difficult to see when the fairy-wrens darted back and forth to the nest entrance.

Watching the nest was far more entertaining than TV. In less than an hour, my tally sheet filled with visits from all four birds, each of which delivered insects at short intervals. The two adults made more visits than the two younger members of the group, but none was slacking. I kept wondering why the younger helpers chose to attend a nest that wasn't their responsibility.

The more I watched, the more I began to visualize the entire group as a family instead of as a collection of individuals. I couldn't make myself think of them as selfish, calculating birds, each one trying to game the others with a different strategy. The thing is, those birds didn't just raise their babies together, they did *everything* as one unit. Whether eating, sleeping, or prowling the edges of their territory, all four members were rarely separated by more than a few yards. When an intruder showed up, they joined in a vocal defense. Sometimes, two of them would sidle up to each other on a branch—usually the dominant male and female, but not always—and tenderly preen each other's feathers for a few minutes. The male and female also often sang duets with each other, a behavior about as rare as cooperative breeding among birds.

There's something to be said about being part of a group, I thought, recalling that my own nearest neighbor was nearly 100 miles away. Though the benefits may be hard to measure, group living must give a certain level of satisfaction to each member, a feeling of belonging, of purpose. In return, each bird must also sacrifice a bit of freedom, just as a couple of single-celled organisms once did to form something more complex. Helping out in chores such as nest duty is part of the deal. You put in what you get out.

A GLASS OF ALTRUISM may be half full or half empty. Scientists and philosophers tend to pick sides, and, though neither is wrong, the camp you choose depends on your worldview.

At least altruism is easy to define. But does the pure thing actually exist? To science, no good deed can be proven to be completely altruistic because there is always the possibility of hidden benefits. You might as well try to confirm or deny the existence of magic. It's really a matter of personal philosophy.

Optimists like to bring up charities as a shining example of human altruism. Anyone who donates to charity, they argue, can't possibly expect to be repaid—such organizations, by definition, benefit only those who have nothing to give in return. Because philanthropists make gifts with no strings attached, they seem to fit the definition.

But then the "half empty" folks step in. First off, they say, most people who make charitable gifts do so at negligible cost to themselves. For instance, Bill Gates gives away 95 percent of his wealth—a laudable effort—which leaves him with, oh, a couple billion left in the bank. He could buy an extra fleet of Ferraris instead, but that would make no difference to the man's ultimate survival or status. He's already made his fortune, so he can afford to give some away.

And those who donate to charity may receive a benefit more potent than currency: Their friends will hear about it. Reputation is a powerful force. As Martin Nowak's model showed, it can even affect decisions as basic as whether or not to cooperate with a stranger. Your reputation spreads beyond your circle of friends and lingers after your death. Most people want to leave a legacy.

Even anonymous gifts may not be truly altruistic. Consider the group of unknown donors that recently decided to cover college tuition for every kid, present and future, who graduates from high school in Kalamazoo, Michigan. Surely, that's as altruistic as it gets—those kids don't even know who to thank for their free ride. But aside from the obvious economic impacts of such a promise on the city, which will inevitably trickle down to its distinguished citizens, those donors can count on a more subtle and immediate benefit: personal satisfaction.

Don't underestimate that warm, fuzzy feeling. In a recent study of giving and its effects on the human brain, subjects were analyzed while they made anonymous decisions about where to direct funds for charity. Scans showed that the mesolimbic pathway—the part of our brain associated with intense pleasures, such as food and drugs—lit up when people decided to give away their money, suggesting that generosity fulfills us on a primal level. In another study, participants were given five dollars to spend by the end of the day either on themselves or on gifts, and those who bought presents for others reported feeling significantly happier than those who spent the cash on themselves. It turns out that money *can* buy happiness, but only if you give it away.

Because satisfaction is a benefit, it technically violates the definition of altruism. But the argument begins to seem curiously circular: Altruism doesn't exist because it makes you happy.

Of course it makes you happy. I watched the story unfold

between those purple-crowned fairy-wrens in the rough Australian outback. They probably got a little spike of pleasure with every positive interaction, whether by bringing extra food to a nest or snuggling up together on a branch for a session of mutual preening. Helpers may further their own genes and inherit better territories, but those very particular benefits are part of a larger system of day-to-day behavior. Fairy-wrens are social creatures, just like us. Their brains probably light up with pleasure every time they do good, the same as ours do.

People who have performed extremely heroic acts, such as diving into a burning building to save a child, invariably report feeling a sense of duty rather than the urge to flaunt their bravery. Any decision to risk your own life to save another's is entirely voluntary and selfless, yet the people who do it don't even think about it—those decisions are based on a larger code. If you think altruism doesn't exist, then you believe that heroes don't, either—that the selfish reward they get from saving a life is calculably greater than the life itself.

With its insistence on measurable quantities, science will probably never get to the bottom of the altruism debate—especially in animals. There's no particular reason why humans should be different in this regard; we have the same capacity (if not more) to bond with one another as fairy-wrens and many other social creatures. Reward and fulfillment exist no matter who you are, but they are tough to measure.

In the nearly 1,000 hours I spent watching fairy-wrens in the outback, I became convinced that if cooperation in fairy-wrens is just a game, then at least it's a game they play because it gives the birds some immediate sense of fulfillment. Their altruistic behavior seems to reflect an overall code of living as much as an evolutionary imperative; fairy-wrens behave generously because that's the kind of bird they are, and everything works out. While you might conclude, from mathematical theory, that their helpful behavior is coldly calculating, this judgment confers our own values on the birds as much as the suggestion that they behave strictly out of love for one another. And the debate misses an ultimate point. Sure, we can never know whether or not real altruism exists in this universe, but wouldn't it be wise for us—considering the bleak alternative—to take a cue from fairy-wrens, and act as if it did?

wandering hearts

THE TRICKY QUESTION OF ALBATROSS LOVE

The romantic life of an albatross is a sweeping kind of romance, the dreamy feeling that all horizons open to an unlimited universe where anything is possible, given sufficient time and space. Albatrosses exist so close to infinity that on a windy day and in the right state of mind, far from the sight of land, the casual observer may be forgiven for forgetting that these are earthly creatures. What would it be like to drift on a pair of six-foot wings, pushed by the breeze, beyond the last vestiges of mundane life?

There's something spiritual about albatrosses. With 95 percent of their lifetime spent over the open ocean, they live so much differently from how we do—from how most creatures on the planet do, actually—that it's hard to conceive of these birds even breathing the same air, which may be why their lifestyle seems so romantic. I have watched people break down and cry upon visiting an albatross colony on a far-flung island, caught off guard, sinking to their knees as if on a religious pilgrimage. Sparrows don't exactly elicit the same response.

Everything you've ever heard about albatrosses is probably true, and then some. We always suspected that they cover vast distances, but never imagined just how far until scientists began attaching global-positioning-system tags to individual birds in the 1980s.

The numbers are staggering. A gray-headed albatross was recently recorded circling the entire Southern Ocean, all the way around Antarctica, in forty-two days, and it kept sailing east with the wind at its tail, completing several round-the-world circuits in the following year. Laysan albatrosses nesting on tropical Pacific islands routinely glide 2,000 miles up to Alaska and back just to grab a quick snack for their hungry chicks. Wandering albatrosses, with the largest wingspan of any flying bird—nearly twelve feet from tip to tip—may log several hundred miles per day, even resting on the wing. Circumstantial evidence suggests that they can shut off half their brain at a time, catching sleep at 40 miles per hour. Because of a special shoulder-locking tendon, albatrosses don't use energy to hold their wings out; their resting heart rate is probably *lower* in flight than it is when they are sitting on the ground. If you multiply the percentage of each day a wandering albatross spends in the air by its average cruising speed and expected lifetime, a typical adult wanderer will have traveled, conservatively, 4 *million* miles in its life—the equivalent of eight round-trips to the moon, more than any other animal on earth, more than any single car humans have ever constructed. Wanderer indeed.

The life of a restless spirit isn't always as romantic as it sounds, however, as any weary traveler can attest. When you're always on the wing, covering a hundred thousand miles a year, it's difficult to sit still long enough to find the daily, earthly romance of a soul mate. One might then logically assume that albatrosses, the rolling stones of the bird world, have made some necessary sacrifices in their love lives. Not so. These globe-trotters, who mate for life and are incredibly faithful to their partners, just might have the most intense love affairs of any animal on our planet. That's what makes these birds so emotionally captivating up close. To see what real devotion is like, you need to spend some quality time with an albatross.

BIOLOGISTS DON'T TALK MUCH about love—the word has too many meanings, is too unscientific for academic use. When discussing love between animals, they tend to use more clinical terms, such as "pair bond" and "monogamous relationship," generally implying attraction without the squishy overtones. There is no way to physically measure love, which is part of what makes it so mysterious and inspiring.

Still, we know a few things about how love works, especially in humans. When two people fall in love, their brains act in predictable, somewhat prosaic ways. To be precise, you never love with all your heart—in fact, you love with your ventral tegmental area and caudate nucleus, nestled deep within your skull where basic drives, like hunger and thirst, are processed. One recent study found that college kids who described themselves as "madly in love" were producing high levels of dopamine, the same chemical released when the brain reacts to cocaine (which gives that fluttering-heart sensation). No wonder lovebirds sometimes do crazy things; they're all high as kites.

Fortunately, those initial effects don't last forever. Levels of dopamine and other chemicals associated with romantic attraction—including pheromones, serotonin, and nerve growth factor protein—seem to drop back to normal within one to three years. Relationships invariably settle down after the first puppy-dog stage; the spark doesn't necessarily go out, but the brain reels itself back into sobriety.

The same flush of chemicals is likely present, at least to some degree, in many animals. But it doesn't tell the whole story. Love is much more than getting high. If that were all, then we'd be running around searching for the next hit—and, with a few exceptions, that's not generally how it works for us. Most people settle into long-term relationships, cozying up with one dedicated partner indefinitely. Biologists explain this as attachment, theorizing that couples stick together for essentially the same reasons that a kid sticks with its mother: The arrangement benefits both parties. Initial lust glues people together and then dries in place. But attachment is harder to measure than lust because it doesn't leave as many traces of chemical activity and can be hard to predict.

The saying "opposites attract" isn't really true when it comes to human relationships. We do often seek out romantic partners with contrasting genes for immune systems, even though we aren't consciously aware of each other's genetic makeup—a tendency that has also been shown in other vertebrates, and likely gives our kids better immunity from diseases. But generally, our ideal partner is a self-image with opposite gender: same social status, same health, same age, et cetera. The more similar you are to your partner, the more likely you are to develop a long-term attachment past the initial head-over-heels stage. The more differences between you, the more cracks will appear over time.

We know instinctively what it feels like to be in a long-term relationship, but it's hard to say whether other animals feel the same way. Do albatrosses get high when they meet a potential mate? What goes through their minds twenty years later, when they're still nesting with the same partner? Do they work through the same stages of love? Some would argue that any expression of emotion in animals is dangerously close to anthropomorphization, but anyone who's ever had a pet, or worked intensively with a few individual birds over a long period of time, can speak to a variety of different moods. Why not outright love?

Personally, I think albatrosses feel love even more intensely than we do, and available evidence seems to back me up. No matter what category of affection you study, albatrosses beat us every time.

ONLY ABOUT 3 PERCENT of the world's 5,000 mammal species are considered socially monogamous. Besides humans, the list includes wolves, beavers, and a surprisingly amorous prairie vole. However, at least 90 percent of all birds pick dedicated mates at some point in their lives. (A few other animals form long-term pair bonds, including the parasitic worm that prefers to copulate with its special lover deep inside the human liver.) Before you jump to rosy conclusions about birds mating for life, though, be warned that appearances are deceptive.

Monogamy among birds isn't a yes-or-no proposition. Relationship strategies span the full range between total devotion and complete promiscuity, and most birds fall somewhere in the middle. For many, loyalty is an arrangement of convenience. In hummingbirds, the bond lasts just long enough to copulate, for perhaps a couple of seconds. Then females buzz off to build a nest and raise chicks on their own while the males sip flower nectar all summer. Most songbirds have summer flings, sticking together barely long enough to raise a brood before going their separate ways. This tends to work out fine because most songbirds don't live very long; it doesn't make sense to commit to a long-term relationship if your partner will probably drop dead within the next year. Accordingly, birds that traditionally mate for life tend to be larger and longer-lived: geese, swans, cranes, owls, parrots, eagles, gulls, penguins . . . and albatrosses.

Even the most dedicated bird couples often sneak around behind each other's backs. Biologists thought for many years that if two birds paired off with each other, they could be defined as monogamous—one male, one female, and their children—but modern DNA testing has shown that this is rarely the case. Most birds do form steady pairs that work together, as it's difficult for single parents to raise young alone, but the offspring from those pairs often have unexpected fathers. In other words, mother birds routinely go for surreptitious quickies.

Some supposedly monogamous birds are more promiscuous than others. The world's loosest bird may be the saltmarsh sparrow, a tiny, mousy inhabitant of muddy coastal wetlands in the eastern United States. One study found that more than 95 percent of all nests included eggs fertilized by other fathers, that the average nest included DNA from 2.5 males, and that the chance of any two nestmates having the same father was a mere 23 percent. Even so, the little sparrows paired off and faithfully raised their broods as if nothing were happening on the side; the only clue to their dalliances was the differing DNA in each egg. Few birds can rival that amount of cheating (though the vasa parrot in Madagascar and the superb fairy-wren in Australia are close), but almost all birds sleep around a bit. The idea of faithfulness among bird couples is mostly a romantic delusion.

Not that female birds don't at least try to stay faithful. A significant percentage of extra-pair copulations in otherwise monogamous birds may be forced by wandering males without the female's consent. Female ducks of several species have occasionally been observed drowning while attempting to evade the advances of aggressive wandering males. But many females sneak out willingly to copulate with males that are not their main partner, sometimes on neighboring territories, with the same result: A father may unknowingly raise chicks that aren't his. Even swans and doves, traditionally the shining examples of true love in the animal kingdom, don't always stay faithful, although they tend to score at the low end of the promiscuity scale. Strict monogamy comes at a price. To be sure of fathering their own chicks, male mourning doves must always stay close to their females and viciously attack intruders, belying their peaceable reputation.

Divorce rates in birds with long-term partners—the percentage of pairs that break up before either bird dies—are instructive. The proportion of divorces varies wildly among birds that are said to mate for life. Flamingos, for instance, are terrible at keeping commitments, with a chart-topping divorce rate of 99 percent. Tropical birds aren't the only adventurers; king penguins are likewise fickle, with a divorce rate near 80 percent (unusually high among penguins, which generally stay together for many years). By contrast, only 5 or 10 percent of swans divorce, and some species of ravens have even lower divorce rates.

Bird divorce is quite well studied, and there are competing hypotheses for why two birds would get together, spend a few years in a relationship, and then go their separate ways. One theory, called the incompatibility hypothesis, argues that some individuals are just not meant for each other even though they might play well with others. Another popular hypothesis says that some slacker birds won't be productive no matter whom they're paired with; mates of these lowlifes should be motivated to leave at the first opportunity.

The latter seems to be true in at least some cases. Female blue tits in Europe generally improve their breeding success after they dump one male for another of better social status, indicating that some males are better than others (dumped males, accordingly, experience lower breeding success). But leaving can also be risky because, without an immediate plan, a sudden divorce puts a bird back into the singles scene, and another mate can be hard to find. One study of gulls discovered that fully a third of female divorcées never nested again, even though some lived another ten years. Sometimes, mates won't leave without a fight; female skuas have been known to kill one another outright in battles over desirable males. You'd think that the males would step in to defend their mates under attack, but they take no part in these disputes—cool and calculating, they know that the strongest female will win. Which makes you wonder: What kind of love lets you stand by and watch your mate be killed, then happily pair up with her murderer as if nothing has happened?

Current predictions estimate that about 40 percent of new human marriages in the U.S. will end in divorce, which places us on about the same romantic level as the Nazca booby, a type of seabird known for, among other things, routinely slaughtering its own siblings inside the nest. It's hard to say how many human couples raise kids with the DNA of only one parent, but it happens. Like birds, we are socially monogamous but don't stay faithful all the time.

And albatrosses? Like us, they are long-lived, and put a lot of effort into raising their offspring. Unlike us, their divorce rate hovers near zero percent, probably the lowest of any bird. Zero percent! Albatrosses make us look like freewheeling swingers. They also have relatively low rates of extra-pair paternity—chicks fathered by a different parent—though in one study researchers found that up to one in five wandering albatross chicks did not match the DNA of its father. Those scientists hypothesized that sneaky mating behavior acts as an escape valve for females trapped in long-term relationships, and helps cut down on inbreeding. For an albatross, divorce is not usually a valid option; splitting up would mean a loss of several years, because it takes a long time for these birds to pick a mate. They have to get it right the first time. Everything an albatross does, it does deliberately—especially when it comes to love.

THE LIFE OF AN ALBATROSS is all about patience. After a wandering albatross hatches, it spends a full nine months sitting alone as a chick in its nest, most of that time in quiet contemplation of its surroundings because it has no siblings. It grows slowly. The chick's mother and father are hardworking absentee parents, combing the distant oceans for food and only occasionally returning to the nest to make a quick dinner delivery. Finally, one day, when the young albatross decides it's ready, it stretches its untested wings and glides out to sea without guidance, to spend the next six years alone, patrolling the most windswept regions of the Southern Ocean. Remarkably, during those first few years of its life, the solitary bird will probably never pass within sight of land.

At about age six, the albatross returns to its home island for the sole purpose of finding a mate. Other adolescent albatrosses also appear from points remote, and things suddenly liven up. After many years at sea without much social interaction, they get together on solid ground—and begin to dance.

The courtship dance of a wandering albatross is complicated, haunting, and humbling to witness. It's very different from the dancing YouTube birds and the boy-band-loving parrots mentioned in an earlier chapter. Two birds face each other, patter their feet to stay close as they move forward and backward, testing each other's reflexes, and point their beaks at the sky. Then, as they simultaneously utter a chilling scream, the albatrosses each extend their wings to show off the full twelve-foot span, facing off while continuing to jockey for position. They touch beaks, throw their heads back again, and scream as if no one were watching—which is usually the case, because the birds nest on only a handful of remote and inhospitable southern islands. The courtship dance can continue for several minutes, with the birds responding to each other's cues like a pair of ballroom pros.

When young albatrosses first begin dancing, they instinctively know the moves but transition awkwardly from one to the next. Practice makes perfect, even in the albatross world. Very young birds group up in gangs of half a dozen, facing inward in a ragged ring like high schoolers at the prom. They watch one another closely, mirroring one another's sequences and gradually improving their own personal style. As the birds smooth out their technique, they focus on a few preferred dance partners, spending more time in smaller groups until, eventually, each albatross narrows the field down to just one other individual, which will become its mate. By that time, it has spent so much time dancing with that specific bird—it can take years to pick the perfect partner—that the pair's sequence of moves is as unique as a lover's fingerprint. If you were to write down exactly how they executed their dance, you would see that each pair does things slightly differently but performs the motions the same every time.

After albatrosses settle down with a lifetime partner, their dancing days are numbered. They perform bits of the sequence to greet each other from time to time, but as the years slip past, each pair spends less time dancing and more time raising their chicks. When they commit, they quit the singles scene and move on to the next stage of their lives.

Between the adolescent years at sea and subsequent years of dancing, a wandering albatross might be fifteen years old by the time it nests for the first time. From then on, it will generally stick faithfully with its mate until one of them dies, which might not be for another fifty years. (Some people believe that albatrosses occasionally reach one hundred, but we don't really know, because nobody has studied them that long. The oldest recorded wild albatrosses have been documented raising chicks past age sixty.) Those years are lived slowly, deliberately, at the pace of an imponderable environment. There are few distractions in the life of an albatross, so the birds concentrate on things that matter most—such as one another.

Although albatrosses form long-term partnerships, the time they spend with their mate is limited. The birds nest at most every other year; because the process of raising a chick takes so long, they can't pull it off every summer. When they're not nesting, the albatrosses patrol the high seas across endless miles and vast expanses of ocean. At sea, pairs don't stick together—it would be too easy to get visually separated, and they'd have to spend too much energy keeping track of each other's whereabouts. So even the most committed partners habitually spend months at a time alone, without knowing what their mates are up to.

Nobody knows how pairs of albatrosses decide when to meet after these long intervals apart. Sometimes they skip a year between nesting seasons, sometimes two. But invariably, each shows up on the nesting island at about the same time, almost as if the date were prearranged. The first meeting is businesslike; if both birds are healthy, they'll get right down to work. Male albatrosses generally gather the nest material; females do the interior decorating. When the female lays an egg, she loses about 10 percent of her body weight—wandering albatross eggs are four inches long and weigh more than a pound. The male takes the first, longest incubation shift while the female goes to sea to replenish lost calories, and then they switch at regular intervals until the egg hatches. Both parents feed the chick until it is old enough to fly away.

The entire nesting process takes about a year, but the parents spend most of that time apart as well. They see each other to mate and build the nest together, but as the season continues, they stay at sea for increasingly long intervals. During incubation of the egg, they must wait for each other to trade places (in one case, when an albatross died at sea, its widowed mate sat on an infertile egg for 108 days before finally giving up). But once the chick can stay warm on its own, each parent comes and goes on its personal schedule and the two rarely encounter each other at the nest.

For most of the year, their relationships are strictly long-distance. Doesn't seem very romantic, does it?

But think again. Albatrosses are world travelers by nature. They're never going to settle down like some farm chicken. Yet they manage to maintain relationships spanning oceans and decades, with little infidelity and virtually no breakups. These birds don't have cell phones to stay in touch; they spend months at a time pursuing a solitary existence at sea, not even knowing whether their partner is still alive, with only hope and expectation that they might meet again on some desolate isle when the time comes. Not many humans could make that situation work. Albatross pairs rely on a nearly unbreakable bond to stay committed through time and space. Do they think fondly of each other while winging so many millions of miles alone? The more I ponder the lifestyle of an albatross, the more fabulously romantic it seems.

The birds certainly make the most of their limited time together at the nest. They often sleep with the head of one bird cozily pillowed against the breast of its mate. Pairs habitually rest side by side, sometimes preening the fine feathers on each other's heads with the tender caresses of the most careful lovers. Different people report seeing various things deep in the inky-black eyes of an albatross—wisdom, serenity, wilderness, peace, endurance—which are well and good, but all I see is love.

MOST PEOPLE PASS their entire lives without seeing a single albatross. Case in point: Samuel Taylor Coleridge. In 1798, he penned his best-known poem, *The Rime of the Ancient Mariner*, which popularized the conception of albatrosses as bad luck—although in the poem the curse arose from the *killing* of an albatross. Coleridge never saw one of the birds in the flesh. He fabricated the story based on mariners' tales, and went to his grave without once gazing upon the bird that made him famous.

Albatrosses live so far from usual human haunts that to encounter one you must step out—way out—of your comfort zone. To see one up close, on its nest, a long journey is required. These days, fortunately, it's a bit easier than it was in the 1700s.

And so it was that I found myself on a ship bound for the Falkland Islands, just northeast of the southern tip of South America. I was working as a staff ornithologist on three consecutive round-trips to Antarctica, back and forth across the Drake Passage from Argentina, on a Finnish-built, Russian-operated research vessel converted for Canadian-organized expedition cruises. Because the Falklands are nearby and offer their own wildlife spectacles, they are often included on Antarctic itineraries.

The islands host one of the two largest albatross populations in the world (the other is at Midway Atoll, in the tropical Pacific). Close to a million black-browed albatrosses nest on the Falklands every year, mostly concentrated in just a couple of dense colonies. From a distance, these colonies resemble spilled salt and pepper on an austere but colorful landscape. There are no native trees to the islands; the albatrosses pile their nests on top of bare rocks near cliff edges, typically surrounded by green slopes of chest-high tussock grass and impenetrable stands of yellow-flowering gorse. Sheep farmers own much of the land but fence off bird colonies to minimize disturbance and encourage visits from cruise ships, which are a prime source of local income. The weather is notoriously temperamental, even in summer, when frequent gales punctuate calm periods. Wind gusts can be powerful enough to blow people around like tumbleweeds, but the albatrosses fly straight into the teeth of such storms. It takes more than a hurricane to faze an albatross.

As our ship eased up to West Point Island, a 3.7-mile strip of rock and vegetation toward the west end of the archipelago, heavy weather was not in evidence. Instead, rare sunshine beamed across a stunning vista. Small groups of crisp black-and-white Magellanic penguins porpoised in the water while endemic Falkland steamer-ducks, bulky, flightless fowl named for their peculiar habit of windmilling to safety like paddleboats when panicked, loafed comfortably on a rocky beach. Inflatable Zodiac boats were craned off an upper deck to ferry everyone ashore for the hike to the albatross colony.

West Point has been settled by sheep farmers since the mid-1800s, and most of its surface is grazed flat. I followed a grassy Land Rover track that contoured gently toward the island's opposite, exposed shoreline, serenaded by red-chested long-tailed meadowlarks while keeping an eye out for striated caracaras, raptorlike birds with a penchant for eating baby penguins and occasionally stealing people's hats. After about a mile of easy walking, the track ended abruptly. I could see why: There wasn't anywhere else to go.

The west side of the island dropped over a rampart of 1,100-foot-high sheer rock walls to a turbulent sea. Looking down, I could see albatrosses gliding beneath me against the blue water; thousands of them were also overhead, on either side, and practically underfoot. Where the Land Rover track ended, the albatross colony began. Most of the birds nested on one densely packed slope too rocky for grass to grow, but some also occupied zones of fenced-off tussock. After pausing to take in the view, I stepped from the track and discovered two albatrosses mating on the other side of a thick clump of grass. Though a mere foot and a half away, they didn't stop on my account; I stood there in amazement and watched them continue without a break for five full minutes, one balanced carefully on the other's back. Up close, the albatrosses were huge—bodies like overstuffed pillows, heads the size of soda cans, webbed feet nearly as large as my hand with fingers at maximum spread. Their plumage, so starkly white-bodied and black-winged at a distance, was subtly beautiful at close range, with a quizzical black mark above the eye and brushed yellow and red tones on the beak. There was no real reason for them to be mating this late in the summer, as it was much too late to be fertilizing an egg. I wondered if they were doing it for fun, or practicing.

All around, albatrosses tended their nests. Most had small chicks, too young to be left alone, that peeped out from underneath their parents' breasts. Some of the adults were feeding their young, curling their huge heads down to ground level and regurgitating a thick, calorie-rich stomach oil of concentrated seafood that the chicks eagerly gobbled. Already the race against starvation had begun. One adult albatross sat on a dried-out, wasted-away chick carcass. I wondered how long it would continue to brood its dead baby before giving up. If an albatross nest fails, the birds won't try again for at least another year or two. It must be hard to let go.

The colony breathed life. Incoming albatrosses, with eight-foot wingspans (black-browed are one of the smallest albatrosses, much more compact than wanderers), maneuvered on the breeze to stall at just the right moment over their territory. Sometimes they misjudged and crash-landed, which made me a bit nervous about the birds swooping just a few inches above my head; the impact of a nine-pound albatross traveling 30 miles per hour could knock a person senseless. One knifed so close that I could feel the whoosh of air against my cheek, and I closed my eyes, enjoying the sensation of being so near to majesty.

Among the albatross nests, keg-shaped rockhopper penguins huddled in rocky crevices, showing off their rock-star hairdos and tuxedo body styling. I wondered what the albatrosses and penguins thought of each other, 70 million years after evolving from a common ancestor. Over those aeons, albatrosses grew long wings and became the greatest fliers in the world while penguins lost their ability to fly and became swimmers instead. They still seemed to share the same capacity for tenderness. Like the albatrosses, mated penguins like to stand close together, bodies touching, and I watched several pairs tenderly preening each other's bristly feathers. Though penguins and albatrosses seem incomparably different, they maintain similar nesting routines, as do most large seabirds. It wasn't hard to see a bit of us in their behavior, either. The birds reminded me of old couples, past the infatuation stage but locked in deep, enduring bonds.

How did those routines develop? Nobody really knows. Darwinian natural selection can often explain why different creatures are well adapted to their respective environments—why, say, an albatross grew such long wings in the windiest places on earth—but what about true love? Emotions don't leave fossils, so we have less hard evidence of their history.

In its usual human sense, romantic love may just be one more survival mechanism to benefit its practitioners—just like the albatross's wing. The strong bonds created by love affairs help us procreate our species. If we didn't fall in love, we wouldn't be as motivated to stick together while raising our children. Human kids take decades of supervision, food, guidance, tuition, and general support. The more help they receive from both parents, the better. If you want to take a purely mechanistic view of love, then it makes sense in terms of survival.

If love has evolved just like any physical characteristic, then there is no reason for it to be uniquely human. All of the same pressures that have acted on us to love each other—long life expectancy, high investment in our children, the necessity of both parents to provide for their kids—have also acted on albatrosses. We're not so different in our basic needs and functions.

IN A FAR CORNER of the West Point colony, I noticed a pair of black-browed albatrosses taking a quiet time out in the midst of all the chaos, seemingly lost in each other's presence. One sat squarely on their nest as a tiny chick's head poked out from beneath its downy breast feathers—flat-out asleep, evidently snoozing after a full meal. The bird's mate snuggled alongside so that their heads rested softly against each other. When one breathed, the other moved slightly. Both had their eyes half closed, completely relaxed. They gave the impression of a pair of lovers leaning against each other on a park bench, gazing into an ocean sunset, mere pinpricks in the folds of our universe, but safe, secure, and content with their place in it. For the moment, nothing else mattered.

ACKNOWLEDGMENTS

In November 2011, I received a four-sentence e-mail that would change my life. It was from Laura Perciasepe, an editor at Riverhead Books. "I'm looking for a writer for a book on birds," she wrote, and thus this project was born. Laura shared her idea for an entertaining book about bird behavior, and then, amazingly, she relinquished control, allowing me to run with the concept over the next year as she cheered, inspired, and applied her deft editing talents. I am completely indebted to her vision, trust, and willingness, when considering possible writers, to call on a twenty-six-year-old bird nerd thousands of miles from Riverhead's editorial offices in New York, and I am deeply grateful to Laura for her unfailing support and encouragement throughout this project. My thanks to everyone else at Riverhead Books, including Geoff Kloske, Kate Stark, Jynne Martin, Katie Freeman, and their teams. A special thanks to Helen Yentus and Janet Hansen for the jacket design. Thank you also to copy editor Amy Brosey, copy chief Linda Rosenberg, managing editor Lisa D'Agostino, production manager John Sharp, interior designer Nicole LaRoche, and all the design and production staff who turned my efforts into this beautiful book. I am likewise indebted to my agent, Russell Galen, of the Scovil Galen Ghosh Literary Agency, who, when I hesitantly contacted him about representing my work, replied, "I've been following you from afar since you were around twelve . . . it's destiny." If our connection was written in the stars, it's now also written in a book. I could not have asked for a more professional, accessible, colorful, and effective agent. This book also greatly benefited from insightful, well-informed, and detailed critiques by two luminaries of the American Birding Association: Ted Floyd, *Birding* magazine editor, and Paul Hess, *Birding* "News and Notes" department editor. Ted and Paul generously reviewed early manuscripts, sometimes on short notice, and their rare combination of ornithological and literary expertise added substantially to this work.

This volume reflects the research of many brilliant scientists, some of whom have employed me in field studies around the world over the past decade. Ornithologists may be a strange lot, but they know about living to the fullest—often in the most remote, spectacular, and adventurous corners of this planet—and these hardcore scientists have taught me about much more than birds. I appreciate all that I have received from those who came before me.

Finally, I am thankful for my parents, whose love and support have helped me follow my own birdy path, and who have embraced the fact that family vacations now inevitably turn into birding holidays. Not all parents can critique a page with the sharp eye of a journalist, argue about bowerbird art over dinner, then bake the best chocolate chip cookies in the world. So thanks, Mom and Dad, for the cookies and for the world.

NOTES AND SOURCES

A world of literature is distilled in this book. Each chapter could easily expand into a volume of its own—and many of them already have been. I don't pretend to cover these subjects exhaustively, but I offer a taste and synthesis of available research to show how cool, interesting, and thought-provoking the world of birds can be. While some of the themes presented are my own interpretations (for instance, that albatrosses can feel love) based on my personal field experiences with birds, the facts are facts. Major sources are listed below in order of appearance within each chapter. Even this is an abbreviated account; instead of including every reference, I have highlighted the most interesting studies and the scientists behind them, pointing the way for anyone interested in further reading.

FLY AWAY HOME

I tracked down the lost racing pigeon's owner, Marty, by looking up its band digits on the American Racing Pigeon Union website (if you ever find a lost pigeon with bands, report it—its owner will thank you!), and interviewed him by phone in the spring of 2012. The subject of bird navigation is huge, and whole books have been written about how birds find their way—for instance, Miyoko Chu's well-researched *Songbird Journeys* (2007) and Scott Weidensaul's excellent exposition of migration, *Living on the Wind* (2000)—so this chapter only hits the highlights of homing behavior. Rosario Mazzeo reported the results of his shearwater experiment in the 1953 article "Homing of the Manx Shearwater." At least three books have been written about Bobbie the Wonder Dog. Ninja was featured on an episode of *Nature* originally aired on PBS in 1999. Yosemite black bear management protocols were outlined in a 1997 California Fish and Game document. Smallmouth black bass homing experiments were described by R. W. Larimore in 1952. Homing behavior of garden snails was described in the 2010 BBC Radio 4 "So You Want to Be a Scientist" winning project, by sixty-nine-year-old grandmother Ruth Brooks, reported by *The Telegraph*. The white-crowned sparrow homing experiments were conducted by Richard Mewaldt in the 1960s and 1970s. Rupert Sheldrake's article "The Unexplained Powers of Animals" was printed in *New Renaissance* in 2003. Andrew Blechman gives a thorough account of pigeon racing history (and other pigeon stories) in his 2007 book, *Pigeons: The Fascinating Saga of the World's Most Revered and Reviled Bird*. Hans Wallraff, who conducted the tilting turntable experiment, reports on the map and compass in his 2005 book, *Avian Navigation: Pigeon Homing as a Paradigm*. Pigeons were tracked following roads by global-positioning-system technology in a 2004 Oxford University study by Tim Guilford and colleagues. The "sun compass" was discovered by Gustav Kramer in his 1951 starling experiments. Stephen Emlen conducted the planetarium tests with indigo buntings in 1967. Mel Kreithen was the first to demonstrate that pigeons can hear infrasound; he investigated their perception of polarized light and pioneered many studies of pigeon navigation. A neural basis for magnetic perception in pigeons was described in a 2012 *Science* article by Le-Qing Wu and J. David Dickman. Katrin Stapput performed the robin experiments showing the right eye's sensitivity to magnetic fields, in 2010. Floriano Papi first proposed an "olfactory map" for pigeons in 1972; olfaction continues to be debated as it relates to navigation. Martin Wikelski published the right-nostril research in 2011. Jon Hagstrum correlated pigeon disappearances with infrasound in 2013. Coverage of the Birdmuda Triangle hit major media in August 2012 after racing pigeon disappearances in northeast England. Airborne pigeon hierarchies were described in a 2010 *Nature* paper. The South African Million Dollar Pigeon Race was moved to the Emperors Palace in 2013 after sixteen years at the Sun City Resort in northern South Africa.

SPONTANEOUS ORDER

If you haven't seen the starling flock video, type "murmuration" into YouTube and be amazed. Richard Barnes has exhibited solo shows of his starling photos (titled *Murmur*) in Seattle, Boston, and New York galleries; Jonathan Rosen wrote the 2008 book *The Life of the Skies*. Jeffrey Goldstein is a professor at Adelphi University, specializing in complexity, emergence, and organizational behavior. Steven Johnson's *Emergence* was published in 2002. Peter Corning's 2002 paper was titled "The Re-emergence of 'Emergence': A Venerable Concept in Search of a Theory." John Conway's Game of Life has spun into an entire field of mathematical research on cellular automata—grids of cells that change by set rules—that continues to yield fascinating insights into physics, biology, and other fields. Craig Reynolds posted a mesmerizing demonstration of his Boids model at red3d.com/cwr/boids (accessed March 2013), worth a peek as it closely mimics the real-life YouTube "Murmuration" video. Predicting the path of celestial bodies in one another's gravitational fields, given only present velocities and directions, is called the "n-body problem," which, so far, has not been solved exactly for more than two objects. The Italian researchers' starling flock model was described in "Empirical Investigation of Starling Flocks: A Benchmark Study in Collective Animal Behavior" (Michele Ballerini et al.), and the topological distance conclusion was reported in "Interaction Ruling Animal Collective Behavior Depends on Topological Rather Than Metric Distance: Evidence from a Field Study" (Michele Ballerini et al.), both papers from 2008. The Shakespeare story is omnipresent in popular accounts of starling introductions, but I know of no accurate primary source. Albert's swarm of locusts, however, is not an exaggerated tale; it was conservatively estimated from qualitative measurements to be a single stream of insects 1,800 miles long, 110 miles wide, and one-quarter to one-half mile deep (!), representing the largest single concentration of animals ever recorded (as described in Jeffrey Lockwood's 2005 book, *Locust*). Starling declines were reported by the Royal Society for the Protection of Birds in 2012. Andrea Cavagna et al.'s 2010 paper "Scale-free Correlations in Starling Flocks" describes the implications of flock correlation lengths, and flocks are discussed in terms of spontaneous magnetization in a 2012 paper, "Statistical Mechanics for Natural Flocks of Birds." George Miller's paper "The Magical Number Seven, Plus or Minus Two: Some Limits of Our Capacity for Processing Information" was originally published in 1956. Andrea Cavagna kindly answered several of my questions by e-mail.

THE BUZZARD'S NOSTRIL

Anecdotes are from my experience attracting turkey vultures to my yard with a deer carcass in June 2000 (inspired by the "Meat-Eaters" episode of David Attenborough's 1998 series *The Life of Birds*). John James Audubon's original account of his vulture experiments was published in 1826 in *The Edinburgh New Philosophical Journal*. Among subsequent rebuttals were three articles by Charles Waterton (1832–1833) in *The Magazine of Natural History*. Darwin's *The Voyage of the Beagle* was first released in 1839 as *Journal and Remarks*. John Bachman's vulture studies were published in a sixteen-page pamphlet (1834) under the title "An Account of Some Experiments Made on the Habits of the Vultures Inhabiting Carolina, the Turkey Buzzard, and the Carrion Crow, Particularly As It Regards the Extraordinary Powers of Smelling, Usually Attributed to Them" (1834). Taxonomy of New World vultures is controversial: Many have argued that they are related to storks (for instance, Charles G. Sibley and Burt L. Monroe, Jr., in 1990), some believe they form their own order, and a recent DNA analysis (Shannon Hackett et al., in *Science*, 2008) suggests that New World vultures are related to raptors. Vulture digestion was described in an engaging 2008 *Audubon* magazine article by T. Edward Nickens. Kenneth Stager included an account of Union Oil workers' stories about turkey vultures with descriptions of his own experiments in his 1964 monograph "The Role of Olfaction in Food Location by the Turkey Vulture (*Cathartes aura*)." The Panama chicken carcass experiment was performed by David Houston and published in 1986. Lab tests of turkey vulture sensitivity to different odors are described by Steven A. Smith and Richard A. Paselk in a 1986 paper titled "Olfactory Sensitivity of the Turkey Vulture (*Cathartes aura*) to Three Carrion-Associated Odorants." Information about avian taste buds is given in Frank Gill's *Ornithology* textbook (2007 edition).

SNOW FLURRIES

The Duluth snowy owl sighting was first reported on the local birding listserv mou-net. Snowy owl totals for the 2011–2012 invasion were based on thousands of reports archived on eBird.com. The *New York*

Times story by Jim Robbins was published on January 22, 2012. I saw the Fern Ridge snowy owl on December 19, 2011. A snowy owl with chicks is painted on the wall of the Cave of the Trois-Frères in southwest France, along with other animals, where figures have been dated to about 13,000 B.C. Illinois birder Rick Remington photographed the encounter between the Chicago snowy owl and peregrine falcon. The 1,000 Washington snowy owls of 1916 are cited in the 2005 book *Birds of Washington*, edited by Terence R. Wahl, Bill Twiet, and Steven G. Mlodinow. The term *superflight* was first used by ornithologist Carl Boch to describe multispecies winter finch irruptions. Victor Shelford's influential 1945 paper was titled "The Relation of Snowy Owl Migration to the Abundance of Collared Lemmings." Studies of irruptive cycles for snowy owls have been mixed; for instance, Ian Newton (2002) reported a mean irruption interval of 3.9 years in eastern North America, but after their statistical analyses Paul Kerlinger et al. (1985) concluded, "We did not find evidence that snowy owl irruptions occur at regular 3- to 4-year periods." The same paper suggested that weather may be a cause of snowy owl irruptions, as other hypotheses don't seem to adequately explain or predict occurrences; nearly three decades later, the question remains unanswered. The Alberta study of snowy owl mortality was described in "Causes of Mortality, Fat Condition, and Weights of Wintering Snowy Owls" (Paul Kerlinger and M. Ross Lein, 1988). The Victoria Island snowy owl chicks were banded by David Parmelee. Karel Voous is quoted from the 1988 book *Owls of the Northern Hemisphere*. Mark Fuller, Denver Holt, and Linda Schueck conducted the pilot satellite tracking study of snowy owls in Barrow, Alaska, between 1999 and 2001. Snowy owls were tracked on sea ice by Marten Stoffel et al. in 2008 ("Long-Distance Migratory Movements and Habitat Selection of Snowy Owls in Nunavut"). Norman Smith works with Mass Audubon on the Logan Airport Snowy Owl Project; maps of satellite-tagged owl movements are posted on the Mass Audubon website (accessed March 2013). The April 1995 *New Internationalist* magazine article about nomads was called "The Facts." The "Out of Africa" theory of modern humans' origin is generally accepted, though dates continue to change; most recently, Fernando Mendez et al. (2013) pushed back the first exodus to 338,000 years ago, exceeding previous estimates. Aki Nikolaidis and Jeremy Gray (2010) examined the DRD4-7R allele's relationship to ADHD disorder; a 2013 paper in *The Journal of Neuroscience* by Deborah Grady et al. linked it to longevity; multiple studies have also linked the allele to novelty-seeking, though others questioned this finding; and its impacts on risk taking were reported in Camelia Kuhnen and Joan Chiao's 2009 paper "Genetic Determinants of Financial Risk Taking."

HUMMINGBIRD WARS

Elizabeth Jones at Costa Rica's Bosque del Río Tigre Sanctuary and Lodge described her hummingbird dilemma during my delightful 2011 visit there and in a subsequent e-mail interview. Paul Kerlinger first popularized the comparison of hummingbird weights to postage stamps. Bee hummingbird measurements are in Felisa Smith's "Body Size, Energetics, and Evolution," in volume 1 of the 2008 *Encyclopedia of Ecology*. Robert C. Lasiewski (1962) estimated the nonstop flying range of a ruby-throated hummingbird at 26 hours and 600 miles based on laboratory-measured calorie consumption. R. S. Miller and Clifton Lee Gass analyzed hummingbird predation and longevity in their 1985 article "Survivorship in Hummingbirds: Is Predation Important?" Eight-year-old broad-tailed hummingbirds were documented by William Calder and S. J. Miller in 1983, and the twelve-year longevity record is listed by the U.S. Geological Survey Patuxent Wildlife Research Center Bird Banding Laboratory. Bat falcon diets were published in 1950 by William Beebe—one of the past century's most colorful naturalists and the subject of an entertaining 2006 biography (he died in 1962). The Nano Drone was described in a February 17, 2011, *Los Angeles Times* article by W. J. Hennigan. Robert C. Lasiewski and R. J. Lasiewski (1967) measured a maximum heart rate of 1,260 beats per minute in a blue-throated hummingbird. The Sierra Nevada hummingbird study was reported by Mark Hixon et al. in 1983. The billion-heartbeat rule is *very* generalized, but it summarizes an interesting trend; Herbert Levine (1997) described the inverse relationship of body size and heart rate, and found a "remarkably constant" lifetime mean of one billion beats in a variety of species. Pace of life was reported in a fascinating 1999 paper by Robert V. Levine and Ara Norenzayan, "The Pace of Life in 31 Countries," and the 2007 study was conducted by Richard Wiseman for his book *Quirkology*. Gerald Mayr described the German hummingbird fossils in *Science* in 2004.

FIGHT OR FLIGHT

Anecdotes are from my three-month 2008–2009 field season with the Penguin Science project, a research collaboration among Oregon State University, PRBO Conservation Science (now known as Point Blue Conservation Science), H. T. Harvey & Associates, the U.S. Antarctic Program, and the National Science Foundation Office of Polar Programs in the Ross Island area of Antarctica. Apsley Cherry-Garrard's quotes are taken from his excellent 1922 memoir, *The Worst Journey in the World*. Charles Darwin's iguana-tossing experiment was chronicled in his 1839 book, *The Voyage of the Beagle*, and inspired David Quammen's insightful essay and 1988 book, *The Flight of the Iguana*. Galápagos National Park regulations and tolerant Galápagos wildlife behaviors are described from my own experiences while living there in 2006. For an example of research using flight distance as an indication of fear thresholds, and a good synthesis of flight distance as it relates to fearfulness in animals, see Theodore Stankowich and Daniel Blumstein's 2005 paper "Fear in Animals: A Meta-analysis and Review of Risk Assessment." Leopard seal attacks were reported in a *National Geographic* news story by James Owen on August 6, 2003. The "tend and befriend" theory is attributed to Shelley Taylor at the University of California, first described in a 2000 *Psychological Review* article and popularized by her 2002 book, *The Tending Instinct*. Robert Plutchik died in 2006 at age seventy-eight after a distinguished academic career, having published eight books and hundreds of articles; the emotional color wheel he developed in 1980 is still used today. Ivan Pavlov won a Nobel Prize in 1904 for his studies of dog saliva, and classical conditioning experiments are often called Pavlovian in his honor. Little Albert remains a well-known case study, though some suggest that John Watson exaggerated his results; for instance, see Ben Harris's 1979 critique, "Whatever Happened to Little Albert?" "Low-road" and "high-road" fear pathways were described by neuroscientist Joseph LeDoux, author of several popular books on the human brain. Swiss neurologist Édouard Claparède performed the 1911 amnesia experiment. The quail study was described in "Mothers' Fear of Human Affects the Emotional

Reactivity of Young in Domestic Japanese Quail" (Aline Bertin and Marie-Annick Richard-Yris, 2004). The New Zealand robin study was described in "Rat-Wise Robins Quickly Lose Fear of Rats When Introduced to a Rat-Free Island" (Ian Jamieson and Karin Ludwig, 2012). Physiologist Paul Ponganis measured 500-meter emperor penguin dives at Cape Washington, Antarctica. The implications of penguin fear, based on Penguin Science project research, were summarized in a 2011 *Science* article by Virginia Morell, "Why Penguins Are Afraid of the Dark."

BEAT GENERATION

Aniruddh Patel was featured in *New York Times* articles on December 14, 2008, and May 31, 2010, and Snowball's story has made the rounds of major media. I learned about manakins deep in the jungles of Tiputini Biodiversity Station in eastern Ecuador, where researchers are focusing on their dancing behaviors. Patel et al.'s paper, "Experimental Evidence for Synchronization to a Musical Beat in a Nonhuman Animal," and Adena Schachner et al.'s paper, "Spontaneous Motor Entrainment to Music in Multiple Vocal Mimicking Species," were published in *Current Biology* in 2009. Patel went on to author a deep, scholarly book called *Music, Language, and the Brain* (2010), which argues that music and language are not independent and should be studied together. Steven Pinker's "auditory cheesecake" hypothesis has attracted heavy criticism; while many evolutionary biologists explain music in terms of natural selection, its survival advantage remains unclear—except that it gives us pleasure (in a recent incarnation of this argument, Henkjan Honing calls music a "game" in his 2011 book, *Musical Cognition*). Human musical evolution is a complex subject; my main point, besides the inspiration of curiosity, is that our music and language may share more with parrots, and other animals, than we realize.

SEEING RED

The world's chickens have been counted by the Food and Agriculture Organization of the United Nations, the Global Livestock Production and Health Atlas, and other groups, though all totals are necessarily estimates. Per capita U.S. meat consumption is closely monitored by the Livestock Marketing Information Center. Thorleif Schjelderup-Ebbe's 1921 dissertation was first translated into English in 1927, according to Porter Perrin, who reviewed the term "pecking order" in 1955. Schjelderup-Ebbe's studies have been referenced by many—for instance in a 1988 Stanford essay, "Dominance Hierarchies," by Paul Ehrlich et al. Colin Allen wrote an in-depth review of transitive inference in animals in a chapter of the 2006 book *Rational Animals?* Joseph Malkevitch investigated tournament graph theory in a featured essay for the American Mathematical Society called "Who Won!" H. G. Landau's theorem was originally published in his 1953 article "On Dominance Relations and the Structure of Animal Societies." Randall Wise's red chicken contacts were reported in *Los Angeles Times* and *New York Times* articles in 1989, as well as in other media.

CACHE MEMORY

The full text of Lewis and Clark's journals, totaling nearly 5,000 pages, is archived at lewisandclarkjournals.unl.edu (accessed March 2013). Lewis and Clark's supply list is on the *National Geographic* website. H. E. Hutchins and R. M. Lanner (1982) documented nutcrackers caching up to 98,000 seeds in one season (often burying several seeds in the same cache). Nelson Dellis has been featured on CNN and Fox News, in *Forbes* and *The New Yorker*, and in other media. Joshua Foer wrote a fascinating and highly readable book about the U.S. Memory Championship called *Moonwalking with Einstein* (2012). Johannes Mallow's records are listed on the World Memory Statistics website (world-memory-statistics.com; accessed March 2013). Stephen Vander Wall's five hypotheses about cache recovery are described in his 1982 paper "An Experimental Analysis of Cache Recovery in Clark's Nutcracker." Vander Wall continues to research caching behavior in various species, wrote the 1990 book *Food Hoarding in Animals*, and is an associate professor at the University of Nevada, Reno. Paul Reber, a psychology professor at Northwestern University, estimated the human brain's capacity at 2.5 petabytes in a *Scientific American* article on April 19, 2010. *Wired* mentioned in a February 2011 article by John Timmer that a single human brain could perform the calculations of all the world's computers combined—a vague but interesting assertion. Martin Hilbert and Priscila López estimated the information contained in a single human's DNA (not the brain) at 30 zettabytes (30,000 exabytes) in a 2011 *Science* paper. The student vs. nutcracker study was mentioned by Richard Cannings in his 2007 book, *The Rockies: A Natural History*. Relationships between hippocampal volume and memory were discussed in a 2004 paper by Cyma Van Petten. Captive chickadee hippocampi were shown to shrink 23 percent in a 2009 Cornell University study by Tim DeVoogd and Bernard Tarr.

MAGPIE IN THE MIRROR

The Eurasian magpie mirror test is reported in Helmut Prior et al.'s 2008 paper "Mirror-Induced Behavior in the Magpie (*Pica pica*): Evidence of Self-Recognition." Mirror tests with young children are described in Beulah Amsterdam's 2004 paper "Mirror Self-Image Reactions Before Age Two." The history of the word *magpie* is given in the *Funk & Wagnalls Wildlife Encyclopedia* (1974). Magpie superstitions are discussed in a 2008 BBC News Magazine article by Denise Winterman, "Why Are Magpies so Often Hated?" Sang-im Lee et al. inferred magpie phylogeny from mitochondrial DNA data in 2003. The paper by Won Young Lee et al. describing magpies recognizing human faces was published in *Animal Cognition* in 2011. The *Manchester Evening News* reported a magpie stealing car keys and tools in 2006, and *The Telegraph* reported a magpie stealing a woman's engagement ring in 2008. Marc Bekoff described the magpie funeral (along with other examples of animal emotion) in his 2009 paper "Animal Emotions, Wild Justice, and Why They Matter: Grieving Magpies, a Pissy Baboon, and Empathetic Elephants." A 2012 *Time* magazine article by Jeffrey Kluger profiled loyalty and friendship in dolphins and other animals, empathy in rats was described in a 2011 *Science* paper by Inbal Ben-Ami Bartal and coauthors, and mourning-like behavior was described in elephants in a 2006 *Biology Letters* paper by Karen McComb and colleagues. Gordon Gallup, Jr.'s pioneering mirror tests with chimpanzees were published in a 1970 *Science* paper. Studies of human self-recognition in patients with autism, schizophrenia,

Alzheimer's disease, and brain injuries are summarized from Gallup et al., "The Mirror Test," in the 2002 book *The Cognitive Animal*. Jens Asendorpf et al. conducted the 1996 study showing children may be influenced by others' behavior on the mirror test. The story of Phineas Gage was related in a 2010 *Smithsonian* magazine article. Michael Benton (1990) gave the figure of 300 million years for divergence of birds and mammals in a *Journal of Molecular Evolution* paper. Robert Epstein et al. trained pigeons to pass the mirror test (without apparent self-recognition) and reported their results in a 1981 *Science* paper.

ARTS AND CRAFTINESS

I observed great bowerbirds during a six-month field season at Mornington Sanctuary in northwest Australia (see the chapter titled "Fairy Helpers"). The great bowerbird with a toy soldier was photographed by Tim Laman and accompanied a 2010 *National Geographic* feature about bowerbirds by Virginia Morell. Theft of colorful wire in great bowerbirds was studied and described by Natalie Doerr (2010). Richard Dawkins, author of the influential 1976 book *The Selfish Gene*, published *The Extended Phenotype* in 1982. John Endler's 2012 paper "Bowerbirds, Art, and Aesthetics" appeared in *Communicative and Integrative Biology*. As he points out, the *Stanford Encyclopedia of Philosophy* has a helpful, in-depth discussion of the definition of art on its website, revised in 2012. John Endler et al.'s 2010 paper about bowerbird forced perspective appeared in *Current Biology*, with a 2012 follow-up by Laura Kelley and John Endler ("Male Great Bowerbirds Create Forced Perspective Illusions with Consistently Different Individual Quality") concluding that this trait varies by individual. Denis Dutton's book *The Art Instinct* was published in 2010. Jared Diamond authored a paper about bowerbirds and the evolution of aesthetics in *Nature* in 1982. His study of bowerbird style (with conclusions about culturally transmitted visual taste) was reported in his 1986 paper "Animal Art: Variation in Bower Decorating Style Among Male Bowerbirds *Amblyornis inornatus*." Odoardo Beccari's quote is taken from a chapter in volume 2 of Samuel Lockwood's *Readings in Natural History* (1888), titled "The Bower Birds—Avian Aesthetics." The chimpanzee painting story is well documented in a Museum of Hoaxes Web article called "Pierre Brassau, Monkey Artist, 1964." Specific bowerbird taxonomy is unclear; these birds are thought to have descended from the crow family along with many other Australian birds, and are now perhaps most closely related to lyrebirds (Charles G. Sibley et al., 1984). Joah Madden described bowerbird brain sizes in his 2001 paper "Sex, Bowers, and Brains."

FAIRY HELPERS

I spent six months at the remote Mornington Sanctuary, a property of the Australian Wildlife Conservancy in interior northwest Australia, in 2010, as part of a multiyear study on purple-crowned fairy-wrens funded by the Max Planck Institute for Ornithology, under the able direction of Michelle Hall. Results of this research were published in a 2012 paper by Sjouke Anne Kingma et al., "Multiple Benefits of Cooperative Breeding in Purple-Crowned Fairy-wrens." I highly recommend Richard Dawkins's classic 1976 book *The Selfish Gene*, which outlines the case for altruism as a means of furthering genetic legacy. The concept of kin selection can be traced to Darwin; more recently, people such as J. B. S. Haldane have computed its effects precisely, and author-scientists such as E. O. Wilson have explained it to the rest of us (see his 1980 book, *Sociobiology*). Game theory is the study of strategy, completely separate from biology, but there are fascinating parallels; John Maynard Smith and other biologists have used game theory to enhance our understanding of evolution, from sex ratios to territoriality and animal communication—and altruism. Mathematicians Merrill Flood and Melvin Dresher originally formulated the prisoner's dilemma in 1950, and political scientist Robert Axelrod reported on his iterated prisoner's dilemma tournament in his 1984 book, *The Evolution of Cooperation*. Stephen Majeski argued that arms races are prisoner's dilemmas in his 1984 paper, "Arms Races as Iterated Prisoner's Dilemma Games." Andrew Russell's 2007 paper about maternal investment in superb fairy-wrens is titled "Reduced Egg Investment Can Conceal Helper Effects in Cooperatively Breeding Birds." Martin Nowak's 2011 book *SuperCooperators* explains his argument that cooperation should be considered a third tenet of evolution. The neuroscience of charity was reported in an article in

The Economist of October 12, 2006, "Altruism: The Joy of Giving." Judith Lichtenberg, a philosophy professor at Georgetown University, wrote an insightful essay titled "Is True Altruism Possible?" in an online *New York Times* opinion page of October 19, 2010.

WANDERING HEARTS

Anecdotes are from the Carcass Island and West Point black-browed albatross colonies in the Falkland Islands in 2012, when I worked as an onboard ornithologist for three cruises to Antarctica with One Ocean Expeditions. I heartily recommend Carl Safina's book *Eye of the Albatross* (2003), which paints a convincingly realistic portrait of life from an albatross's perspective. Gray-headed albatrosses were documented circumnavigating Antarctica in a 2005 *Science* paper by John Croxall et al. Brain scans of love-struck college students were analyzed by Andreas Bartels et al. (2001), who described three stages of love: lust, infatuation, and enduring love. Social monogamy rates in mammals and birds are given in the textbook *Animal Behavior* by John Alcock (ninth edition, 2009). Sexual monogamy rates of saltmarsh sparrows were measured by Chris Elphick et al. in 2009. The topic of bird divorce was explored by Susan Milius in an engaging 1998 *Science News* article. Divorce rates have been studied in New Zealand's red-billed gulls by James Mills. André Dhondt and Frank Adriaensen reported blue tit divorces in "Causes and Effects of Divorce in the Blue Tit *Parus caeruleus*" (1994). Human divorce rates are the subject of much quibbling, but overall trends are definite: In the United States, divorce rates tripled between 1950 and 1980, then flattened and declined, and about 40 percent of today's new marriages are projected to end in divorce. David Anderson has studied divorce in Nazca boobies, reporting a 38 percent annual rate ("Serial Monogamy and Sex Ratio Bias in Nazca Boobies," 2007). Pierre Jouventin et al. reported a 0.3 percent divorce rate in wandering albatrosses from the Crozet Islands in a 1999 *Animal Behavior* paper. Genevieve Jones documented an 18 percent extra-pair paternity rate (illustrating the difference between social and sexual monogamy) in wandering albatrosses in 2012, after an earlier genetic study had estimated a 10 percent extra-pair paternity rate in 2006. The oldest known albatross is a female

Laysan albatross named Wisdom, who, as of 2013, was still raising chicks at age sixty-two; John Cooper et al. reported a wandering albatross estimated by band recovery to have been at least a half-century old in 2001, noting that "demographic studies need to continue for several more decades" before we learn how long they might live in the wild. Martin Gardner, in *The Annotated Ancient Mariner*, points out that Samuel Taylor Coleridge probably didn't realize just how big an albatross is; a flopping twenty-pound bird with a twelve-foot wingspan wouldn't so much drape around a sailor's neck as drag heavily on the ground. Not all metaphors should be taken literally.

INDEX